HOW MANY SOCKS
MAKE A PAIR?

HOW MANY SOCKS MAKE A PAIR?

Surprisingly Interesting Everyday Maths

BOOKS

First published in Great Britain in 2008 by
JR Books, 10 Greenland Street, London NW1 0ND
www.jrbooks.com

A catalogue record for this book is available from the British Library.

ISBN 978-1-906217-59-4

1 3 5 7 9 10 8 6 4 2

Printed by MPG Books, Bodmin, Cornwall

Contents

Acknowledgements

Over the years so many people have passed on so many mathematical nuggets to me that sometimes it is hard to recall where I first encountered the idea. Even when I can remember who introduced me to a particular idea – such as the prisoner puzzle in chapter 5 (thank you, William Hartston!) – the inventor seems to have been lost in the mists of time.

There are, however, a number of people whom I can acknowledge directly. Top of the list is Martin Gardner, whose columns in *Scientific American* were an inspiration to me and many of my generation. It was his books that first drew my attention to Dragon curves, Gilbreath shuffles and so much more besides.

Two personal websites have been important references: Ron Knott's pages on *Fibonacci*, and Jason Doucette's reports of palindrome records.

I'd like to thank David Wells for his thoughts on triangles, Art Benjamin for tips on squaring numbers, Barry Lewis for his insights on Lucas, David Singmaster for a couple of stunning card tricks, Liz Meenan for her envelope tetrahedron, and John Webb for first showing me the mirror problem.

I've also had help from John Haigh, Colin Wright, Tony Mann, David Bellhouse, Tim Baxter and dozens of others, including a helpful staff member at *Fur & Feather* magazine who educated me on the breeding habits of rabbits.

For the umpteenth time, Richard Harris has been the perfect sounding board, and his comments have had a significant impact on the shape of this book.

Thanks to Lesley, Jeremy and the rest of the JR team for

being so good to work with, and to Graham for putting a sock in it (as it were).

Erica, it was a breath of fresh air to have such an enthusiastic, inquisitive and receptive student. Greenwich Park is a good location for a maths lesson.

Thank you Elaine and Barbara for going through the manuscript in detail and for all your invaluable suggestions.

Finally, I'd like to thank Mary Harris and Poppy Brech who, as self-confessed maths-phobes, gave me some priceless insights into what NOT to put into a book like this one.

Introduction

How Many Socks make a Pair?

The answer isn't two. Not in my house, anyway. Why not? Because I can guarantee that on a dark winter morning, when I reach into my drawer of mixed black and blue socks and pull out two, they are invariably odd.

The good news though is that however unlucky I am, if I take THREE socks out of my drawer, I am guaranteed to get a pair. It might be two black or two blue, but there will certainly be two of something. So it only takes one extra sock for the power of mathematical law to overcome Murphy's Law. How many socks make a pair? Three, if you want to be certain of it.

When there are two types of sock, remove any three
to guarantee a pair.

That's only true if there are only two colours of sock, of course.

If there are three types of sock in the drawer – blue, black and white for example – you need to take out four for a pair. With 10 types of sock, you need to take out 11. And in mathematical shorthand, if you have N types of sock, you need to remove N + 1 socks to guarantee a pair. (I promise not to mention 'N' again.)

I like the sock problem, because within its practical every-dayness, there is a lovely mathematical idea that tickles the imagination, and takes the idea of maths beyond the functional (but rather dull) world of $1 + 1 = 2$. It belongs to a world of maths that is surprisingly interesting – even for those people who are sworn haters of all things mathematical.

This is a book about mathematical ideas that anyone can enjoy. It was inspired by a phone call I received one morning from a national newspaper. Erica Wagner, the literary editor of *The Times,* wanted to know more about maths, and thought that I might be able to help.

Her curiosity had grown from conversations with friends, some of whom appeared to have an unfathomable passion for mathematics. She had heard mathematicians describe their subject as 'elegant' and 'beautiful', words often used for poetry or works of art, but... maths? She just didn't get it, and what's more, after those friends tried to explain, she still didn't get it. It's no argument to claim that maths is beautiful because it just... is.

My brief sounded simple enough. A three-hour tutorial to demonstrate what it is about maths that could possibly be described as beautiful. The more I thought about it, however, the tougher the challenge seemed to be. Algebra? Geometry? Calculus? For most people, the only emotions that these words provoke are fear, nausea, or a sort of eye-glazed boredom. Sometimes all three. (This reaction can usually be traced back to a bad experience at school, somewhere between the age of 12 and 16.)

Maths can often make intelligent people feel stupid, or even a little angry. Listening to a mathematician explain something can provoke private thoughts in the listener like this:
'You know, I think this is supposed to be obvious, but I just don't get it.' Or: 'Right now, the main thought going through my head is – **WHO CARES??'**

So when we met for our tutorial, we didn't talk about maths at all. We talked about card tricks. And about mind reading. And limericks. And curious patterns that can appear on a calculator display, apparently from nowhere.

Actually, it's not true to say we didn't talk about maths, since all of the things we discussed were directly connected to mathematics, we just avoided using that word. The biggest problem with maths is the very word itself. It has so many negative connotations that the merest hint that something involves maths is enough to send many intelligent people running from the room.

Our meeting in 2006 was just a conversation, but it sowed an idea, and two years on this book is the result. This is my slightly longer answer to the old question: 'Is it true that maths can be interesting, creative and beautiful?'

In writing it, I am acutely conscious that beauty is a very subjective thing. Just because I find something interesting, or creative, or beautiful does not guarantee that you will feel the same way. In fact just about the only thing I can guarantee is that at some points in this book, whatever your mathematical ability, you are going to be having those very two thoughts (*I don't get it* and *WHO CARES??*) that I hoped you wouldn't.

When those thoughts do arise, please just skip on to the next bit and take comfort from the fact that it's not your fault, it's mine. And somewhere in here I hope that our minds might meet, and you will discover a side to mathematics that you never knew was there.

In writing a book for people who would describe themselves as non-mathematicians, I know that I have often over-simplified, lacked rigour, and often stopped just when (for mathematicians) the maths gets really interesting. That is my little *Apology to Mathematicians*, in homage to G.H. Hardy's book *A Mathematician's Apology*.

Since I have used the word creative to define some of the maths in this book, I'd better define what I mean by it. Back in the 1960s, Arthur Koestler wrote a book called *The Act of*

Creation. In it, he attempted to define what creativity is, and how it happens. He decided that creativity manifests itself in three ways:

Beauty

Discovery

and **Humour**

Later, some anonymous wit thought of a neat way of summarising those three qualities as:

AH

AHA!

and **HAHA**

This book is about the Ah, Aha! and Haha of mathematics.

I don't believe it

You will need: a bathroom mirror, a newspaper, a group of people and some imagination.

Some call it common sense. Others call it intuition, or just 'having a hunch'. Whatever you want to call it, everyone has it to a greater or lesser degree. It's what helps us to understand the world without having to spend too long trying to work things out. And most of the time our intuition serves us very well. But there's a phenomenon that is much loved by mathematicians and scientists, and they call it the counter-intuitive. I have to confess that it is the counter-intuitive bits of maths that I love the best, particularly the examples that seem to clash with our own experience of the world.

In this opening chapter, I want to concentrate on four of my all time favourite examples of the counter-intuitive, each with their own share of Ah, Aha! and Haha.

Let's see how your intuition works.

The mirror

Imagine it's early morning, you are in the bathroom, have just cleaned your teeth and are now inspecting yourself in the mirror. From where you are standing close up to the sink, you can see about half your body – suppose the lowest point that you can see down to in the mirror is your navel.

You now start to step back from the mirror. As you retreat, are you able to see?:

(a) Less of yourself
(b) The same amount
(c) More of yourself.

Think about it for a moment. The answer should be obvious enough, after all you've probably inspected yourself in a mirror thousands of times.

Do you think the answer is (c), that as you step back from the mirror you see more of yourself? That is what most people say. It seems plausible, after all when you try on new clothes in a shop it is normal to step back to get a better view.

It is, however, the wrong answer.

The second most popular answer is (a). As you step back from a mirror you get smaller, so in that sense you see 'less of yourself', though it doesn't answer the question about whether you can now see above or below your navel.

In fact (a) is wrong too. The correct answer is (b) – as you step back from a mirror, the amount of yourself that you can see does not change. This answer comes as a surprise to most people. In fact, it is such a surprise, that it is common for any-one who hears it for the first time to immediately disappear to the bathroom to check it out. You might be tempted to do that yourself right now.

Strictly speaking the answer is only true if the mirror is ver-tical and the floor is horizontal (which is true of most mirrors and floors, but not the tilting mirrors in shops for example). And of course, you mustn't cheat by leaning over when you get close to the mirror (I've known protestors to lean over and shout 'I can see my toes now!').

The mirror question is a lovely example of something that is counter-intuitive. All our common sense points in one direction, yet in this case it fails us. That feeling of a surprise that you want to check out for yourself is something that can be found through-out mathematics. And having checked it out, for many people that is enough. But for others, maybe you, there is a nagging question: why? And it is when you ask why that you are really 'doing maths'.

In the case of the mirror, the 'why' is easy enough to explain. When you stand close to a mirror, what you are seeing is a reflec-tion of light rays from your body bouncing off the mirror and into your eyes. Here's a diagram showing you close up to the mirror:

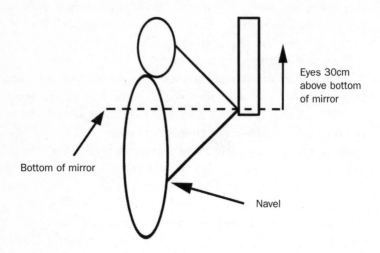

Suppose your eyes are 30cm higher than the bottom of the mirror. If you are standing up straight, this means that the lowest point of your body that you can see will be 30cm below the bottom of the mirror. (This is because a light ray bounces off a mirror at the same angle as it meets it.) As you step back from the mirror, the amount that your eyes are above the bottom line of the mirror doesn't change, it is still 30cm. So however far back you stand, you will always see down to your navel.

That, very informally, is the mathematical proof of the mirror question. It may not convince you, it may not interest you, it may still seem to confound your intuition, but it is right. And you can test it, and talk about it. That, to me, is the best sort of maths.

The wedding party

Now that your intuition has been challenged, let's take another everyday example. Imagine you are at a wedding, and there are about 50 guests, young and old. Chatting idly with other guests, somebody says: 'I wonder what the chance is that there are two people in this wedding party who have the same birthday as each other – not the same birth date, just the same birthday, like 5 May or something.'

You know that there are 365 days in the year. So what is your hunch telling you about the chance of discovering a birthday coincidence in this wedding party of 50 people?

(a) No surprise at all, you think it is almost a dead-
   cert that two of the people share a birthday.
(b) It feels like it's probably about a '50–50' chance.
(c) You would be very surprised. This would only
   happen at, say, one wedding in 7.
(d) You would be freaked out. The chance of this
   must be about the same as that of a rank outsider
   winning a horse race.

All of us have enough experience of coincidences that happen in our lives to build up some sort of intuition for how likely something is. And yet when it comes to coincidences, our intuition is often way off the mark. If you ask a typical group of adults the birthday question, the majority would be surprised by a coincidence. Most answer either (c) or (d). Some will have heard about this puzzle before and will remember that 50–50 features in it somewhere, so plump for (b). In fact the correct answer is (a) – with fifty people in a room, it is indeed almost a dead-certainty that there are at least two people in the room who share a birthday. Assuming that birthdays are evenly spread across the calendar, then the chance of a coincidence is about 97 per cent, so you might expect to go to 30 gatherings of this size and only find one where no birthday coincidence happened.

If you want the chance of a birthday coincidence occurring to be 50–50, you only need to have a group of 23 people. The mathematician Robert Matthews thought of a neat way of identifying a 'random' group of 23 people: just take the two teams in any football match, and add in the referee. If you check the birthdays of the players plus the referee in a weekend of fixtures, then you will find that in half the games, two people on the pitch share a birthday.

And in fact the odds are even better than I've said, because birthdays are not evenly spread around the year. Some months are more fertile than others. Typically there are more births in September than in other months (which gives a hint of what must go on at some of those Christmas parties). This clustering increases the chance of a coincidence by another couple of per cent.

Part of the reason why people are so surprised by this result is that we get confused between the chance of two *particular* people having the same birthday (which is a remote 1 in 365) and the chance of any two people sharing a birthday. In a group of 50 people, there are over a thousand different pairs you could choose from the group – Alf and Bert, Alf and Celia, Bert and Celia ... and so on – so perhaps it's not quite so surprising that among all those combinations you find one where the pair go snap.

If you like this sort of thing, you can actually plot out the chance of a birthday coincidence for a given number of people.

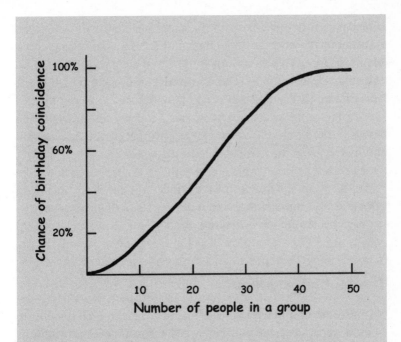

With small groups, the chance of a birthday coincidence grows slowly. With 10 people, it is just 12 per cent, but the curve steepens and doubling the number of people to 20 more than trebles the coincidence chance to 43 per cent. By 30 people it is up to 70 per cent, and at 50 people, when the curve starts to flatten, the chance of a birthday coincidence is up to 97 per cent. Yet you can't be 100 per cent certain of a coincidence until there are 366 people in the room (because one might have a 29 February birthday).

But why do you need just 23 people to get a 50–50 chance of a coincidence? How do you actually work it out? There is a difficult way and a (relatively) easy way. The difficult way would be to say: a coincidence means there might be two people with the same birthday, or three, or two different pairs with shared birthdays, or five people with the same birthday plus a separate threesome. To work out the chance of a coincidence, you could list these (literally) thousands of different scenarios and add up the chance of each one. The alternative approach, which is the one that statisticians always use, is to say: what is the chance that in the group of 23, everyone has a *different* birthday? When we've worked that out, whatever is left must be the chance that there is at least one coincidence in the room.

The chance of a coincidence with 23 people turns out to be $365/365$ x $364/365$ x $363/365$ and so on until you get down to $343/365$, which works out at 0.493, which is near enough 50 per cent.

Do you feel better knowing that? If not, don't worry, we'll move on.

The football stadium

Frimpton United have won the league, and the club decides to hold a reception for the team. The groundsman brightens things up by buying some bunting – flags on a string – to go along the touchline. The idea is that as the players come out of the dressing rooms at the centre of the stand, they will run under the bunting and onto the field. The football pitch is 100 metres long, and the groundsman buys bunting that is 101 metres long. He ties the bunting to the base of the two corner flags on that side of the pitch. The manager spots what he is doing and challenges him. 'You've only got one metre of slack in the bunting!' he exclaims. 'If you lift up the bunting in the middle, is there going to be enough room for the players to get underneath?'

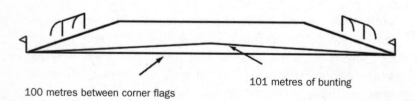

100 metres between corner flags   101 metres of bunting

What happens when the bunting is lifted?

(a) The bunting will be so tight, the players will barely be able to squeeze underneath.

(b) The players will be able to get under the bunting if they crouch very low.

(c) Most of the players will be able to walk underneath without needing to stoop at all.

By now, of course, you smell a rat. This is a chapter about surprises, in which case whichever answer you go for, it's bound to be the other one. So in this three card trick, which do you go for, (a), (b) or (c)? There's one metre of slack, but it has been stretched over a very long distance, so your intuition might well pull you towards (a), that the string can hardly be lifted off the ground at all. But since there is a catch, you go for (b) or (c) instead.

In this case, all three answers are wrong. Not only will all the players be able to comfortably walk under the bunting, but if they get into the team coach, the whole coach can comfortably be driven under the lifted bunting with room to spare. With just a metre of slack in the bunting, the string can be lifted up to a height of 5 metres at the centre of the pitch.

Maybe in this situation, intuition goes wrong because 'sharing' one metre of string along a very long length sounds like sharing five loaves between five thousand people – in other words, you don't expect it to go very far. But that happens not to be the way that the geometry in this problem works. The explanation here is to do with the string forming a very long, thin triangle, and the numbers can be worked out using the ancient theorem of Pythagoras.

Imagine yourself standing at the halfway line. The bunting string and the touchline form two sides of a very long, thin right-angled triangle, like this:

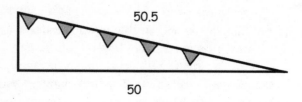

The base of the triangle is 50 metres, and the long side (the hypotenuse) is 50.5 metres. We need to know how high the triangle is, since that's the space through which the players will be passing. You might remember from your youth a rule that says that on a right-angled triangle, the square of the hypotenuse is equal to the sum of the squares of the other two sides. It's called the theorem of Pythagoras. Using Pythagoras, it turns out that the height of the triangle is just over 7 metres, or 23 feet, which is considerably taller than a typical double-decker bus.

But again, it's not so much the maths that interests me here, it's the sheer surprise in the answer.

The newspaper statistics

Let's finish with one more everyday example. Indeed, this one is so 'everyday' that it is perhaps the biggest surprise of all.

Tomorrow's newspapers will be filled with stories, many of them involving numbers, and all of them unpredictable. There may be news of job losses, a company may announce record profits, there may be a report about a number of people killed in a natural disaster, and who knows, perhaps news will come in of a record being broken, such as the largest ever number of toothpicks used to build a model of the Taj Mahal.

But the point is, tomorrow's news is *unpredictable*. Given that these numbers are unpredictable, and that they have no connection to each other, it is hard to imagine that there could be any kind of 'pattern' connecting those numbers. And yet, it is very likely that the numbers *will* form into a pattern.

Pick up any newspaper and look for stories that contain numbers. The sorts of numbers to look for are measurements (height, distance, weight and so on), sums of money (profit, pay, prizes) and statistics that are counting things (number of cars in a pile-up, number of tourists entering the country and so on). You should ignore dates, percentages, people's ages and telephone numbers, for reasons that will be explained shortly.

Find 10 numbers from different parts of the newspaper, and write them down.

Since these numbers are effectively random, the first digit of each number will be 1, 2, 3, 4, 5, 6, 7, 8 or 9. And you might suppose that a number, such as (say) the total visitors at Disneyland last month, is just as likely to begin with a 1 as with a 9. But that is not the case. In fact, the number visiting a randomly selected theme park like Disneyland[1] on any particular day, is extremely likely to begin with a 1, a 2 or a 3, and extremely unlikely to begin with an 8 or a 9. Of the statistics that you have selected at random from today's newspaper, about half will begin with a 1 or a 2, and I'm prepared to bet that you didn't pick any that began with 8 or 9. (You will find them, but they are very rare.)

This remarkable phenomenon is known as Benford's Law. It applies to any selection of statistics, so long as the numbers being chosen are counting something that could be as low as zero, and have no obvious upper limit. Redundancies, numbers injured and sales figures can be just about any number, and generally obey Benford's Law.

However, the ages of people in the obituary column don't obey Benford's Law. The vast majority of people listed in the obituaries are aged between 60 and 90 because (a) the older you are the more likely you are to have done something worthy of an obituary in your life, and (b) because typically the vast majority of the people who die on any given day are aged between 60 and 90. Ages, particularly obituary ages, do not obey Benford's Law. Nor do other constrained numbers including percentages (these will tend to be uniformly spread between 0 and 100) and telephone numbers (these are labels rather than numbers, and do not obey normal statistical laws – for example in London most numbers begin with an 7 or 8).

As with all of the other counter-intuitive examples in this chapter, Benford's Law invites the question: why? How does a mountain 'know' that its height should begin with a 1 far more often than a 9? And even more spookily, how come this works whether we decide to measure the height of the mountain in

---

[1] I should be careful to point out here that I've picked out Disneyland at random, I know nothing about the attendances at Disneyland, Alton Towers or any other theme park.

centimetres, feet or broomsticks? How does an earthquake 'know' that the number of victims will follow a similar pattern?

This is when the power of mathematics can begin to really get under your skin, because it turns out that fundamental mathematical patterns are at work that dictate that Benford's Law will be so. The detailed explanation as to why Benford's Law works is rather involved, and I have no intention of going into it here. But I should at least give a hint of where it comes from. It is related to the fact that the world is full of things that grow 'exponentially'.

A simple example of exponential growth is if you take the number 1 and keep doubling it. The first 25 numbers you get when doubling are:

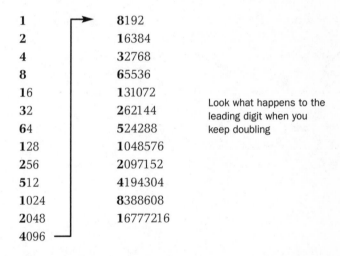

1	**8**192
2	**1**6384
4	**3**2768
8	**6**5536
16	**1**31072
32	**2**62144
64	**5**24288
128	**1**048576
256	**2**097152
512	**4**194304
1024	**8**388608
2048	**1**6777216
4096	

Look what happens to the leading digit when you keep doubling

Notice anything about the leading digits? So far, there have been eight leading 1s, but we have yet to encounter any 9s. Part of the reason why this is the case is that in order to get a leading digit of 9, the previous number that you are doubling has to begin 45..., 46..., 47..., 48..., or 49... . (If you don't believe me, check it out on a calculator). On the other hand, if you double any number that begins with 5, 6, 7, 8 or 9, the answer begins with a 1.

So Benford's Law works for numbers that are forever doubled, and in fact it is also true for numbers that grow by any other fixed proportion, otherwise known as exponential growth.

You've just seen that things as ordinary as a mirror, a birthday, a piece of string and a newspaper can contain jaw-dropping surprises. And all of them can be completely explained by maths. I hope it has aroused your curiosity.

Calculating without a calculator

You will need: your fingers and thumbs (preferably 10).

It might strike you as odd that we have made it to the second chapter, and you haven't had to do any sums yet. After all, isn't that what maths is all about, sums, times tables and such? That's certainly what most members of the public think maths is. And for all sorts of reasons, we aren't as good at sums as we used to be. In 1998 this became a matter of national debate when the British education minister Stephen Byers was asked live on radio to multiply seven times eight and famously gave the answer 54.

Some people blame the world's maths problems on the arrival of the calculator, and it's true, calculators have probably made many of us sloppy when it comes to mental arithmetic. But maths is more than just arithmetic, and calculators haven't just made people lazy about manipulating numbers, they have also taken away some of the creativity that is to be had in discovering how numbers work.

For this chapter, therefore, calculators are banned, and for doing the multiplications that follow the only tool that is permitted is your fingers. That's what the ancients used – it's not a coincidence that they are also called your digits.

Let's start with some times tables. One of the easiest to learn is the nine times table, for which there is a little trick that is well known amongst children, but rather less well known among those who are older.

To do your nine times multiplications, simply put your hands on your knees like this, numbering your fingers left to right as shown:

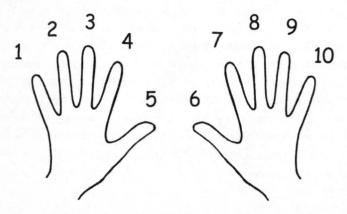

Now choose which multiple of 9 you want. Suppose it's 7 x 9. Simply bend down finger number 7, like this:

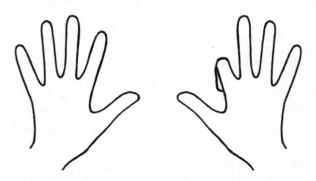

Now count the number of fingers and thumbs to the left of the bent finger (6) and the number to the right (3), put them together and you have the answer: **63**. It works for all 10 fingers and thumbs, so you can quickly check for yourself that 3 x 9 = 27, and 9 x 9 = 81.

You will notice, by the way, that since you are bending over one finger and counting the other nine, that this must mean that the digits of all multiples of 9 between 1 and 10 always ADD UP to 9. (2 + 7 = 9, 3 + 6 = 9 and so on.)

As it happens, if you take any multiple of 9 and add up its digits, then those digits will add to 9, or a multiple of 9. So, for example:

9 x 146 = 1,314, and **1 + 3 + 1 + 4 = 9.**

9 x 542 = 4,878, and **4 + 8 + 7 + 8 = 27,** and **2 + 7 = 9.**

Because multiples of 9 always have this property, you can use it to check that you have done a multiplication correctly. If you multiply 9 x 378 and get the answer 3,502, you must have made a mistake, because 3 + 5 + 0 + 2 = 10. (The correct answer is 3,402).

Will it divide?

How can you tell if a large number is going to be exactly divisible by the numbers 2 to 12, if you don't have a calculator to hand? There are divisibility shortcuts that can answer the question quickly, without you needing to do the whole sum.

Take as an example the number 941,065,236 – which is one digit too large for many calculators to handle. Here's how to check if it is divisible by....

Divisor	Divisibility Test	Number tested 941,065,236
2	If the number is even, then it is divisible by 2.	This number is even, so divides by 2.
3	Add up all the digits of the number. If the sum of the digits is divisible by 3, then so is the original number.	9+4+1+0+6+5+2+3+6 = 36, which is a multiple of 3, so 941,065,236 is divisible by 3.
4	Check the number formed by the last two digits. If it is divisible by 4, then so is the whole number.	The last two digits are 36, which is divisible by 4, hence the whole number is, too.
5	If the last digit is 0 or 5, then the number is divisible by 5.	The last digit is 6, so it fails the 5 test.
6	If the number passes the test for 3 and is even, then it is divisible by 6.	It is divisible by 3 (see above) and is even, so the whole number divides by 6.

8	Check the number formed by the last three digits. If this number is divisible by 8 so is the whole number.	The last three digits are 236, and 236 is not divisible by 8, so it fails the 8 test.
9	Add up all the digits of the number. If the sum is divisible by 9, then so is the original number.	The digits add to 36, which is a multiple of 9, hence so is the whole number.
10	If the final digit is zero, the number divides by 10.	Final digit isn't a zero.
11	Add up the 1st/3rd/5th etc digits, then add up the 2nd/4th/6th digits. If the difference between the two totals is zero, or a multiple of 11, then the whole number divides by 11.	The 'odd' digits are: 9+1+6+2+6=24 The 'even' digits are: 4+0+5+3=12 The difference is 12, so the whole number fails the 11 test.
12	If the number passes the test for 3 and the test for 4, then it is divisible by 12.	It is divisible by 3 and 4 (see above) and so it is divisible by 12.

Notice that the number 7 is missing from these tests. There are some ways of testing for divisibility by 7, but these are not nearly as simple as for other numbers, and it's usually quicker just to do the whole calculation longhand.

Doing your tables

What about those other tricky multiplications, such as the one that the government minister found so difficult, seven times eight.

There's a neat finger technique that works for most of the

common times tables. It is sometimes called 'The Gypsy Method' (it goes by other names too). To use the method, you need to hold your hands in front of your face, with your palms towards you and thumbs pointing upwards.

Call your thumbs 6, and number your other fingers 7, 8, 9 down to your little fingers at 10, as in the diagram. You are now ready to multiply.

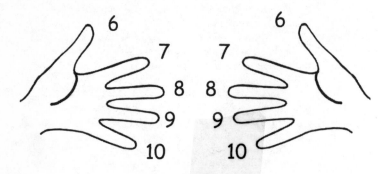

To multiply seven by eight, join finger number 7 on one hand to finger number 8 on the other – like this:

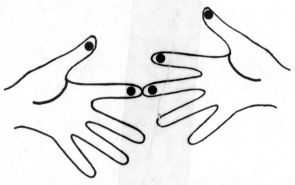

Let's call the fingers that are joined together and any that are above them your Blob fingers (they are marked with a blob on the diagram). And call the fingers that are dangling down the Dangly fingers.

To do the sum, you need to:

- Count up the Blob fingers (there are 5)
- Multiply together the Dangly fingers (3 on the left, 2 on the right, 3 x 2 = **6**)

Now put the Blob and Dangly numbers together, **5 6** and you have your answer, **56!**

It's exciting, isn't it! Want to try another? How about eight times six? You should find that you have 4 Blob fingers and 4 and 2 Danglys, making 4 x 2 = **8** giving the answer **48**.

You have to be careful, because this technique has its limitations. If you use it to multiply 6 x 7 for example, you'll find it gives you the answer 3 12, suggesting that the answer is 312. What's going on? The reason for this apparent error is that the 3 actually represents the number 30, and so you need to add 30 and 12 together, to give the correct answer 42. Fun though this technique is, you might find that simply memorising your tables is ultimately the best method!

I have assumed here that you have the regulation 10 fingers and thumbs. That's not the case for many people – some have fewer, others (like Anne Boleyn according to legend) have more. Fortunately, however many fingers you have the method still works, but if you happen to have an irregular number then you need to use a number system that is based on your number of fingers. Our conventional number system works in 'Base 10', which came about simply because it matched the number of fingers on our hands. So the number 56 means 5 tens and 6 units. It's so obvious you may not have given it a second thought for years. Had we instead evolved to have 12 fingers, we would instead be working today in base 12, so the number 56 (base 12) means 5 twelves and 6 units, i.e. 66 in Base 10.

So how does the finger method work if you have an unconventional number of fingers?

The secret is to start the numbering at your little (or end) fingers, and number the other fingers from there. So if you have eleven fingers, call your little fingers 11, but if you have nine then call the little fingers 9. Here's how an eleven-fingered person would do their times tables.

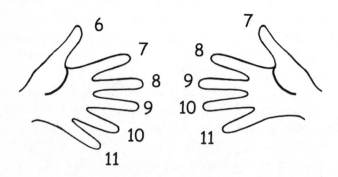

Suppose Anne Boleyn wanted to multiply 10 x 10. Using the eleven-finger diagram you can check that the answer comes to 9 Blobs and 1 x 1 = 1 Danglys, so the answer is 91. But shouldn't the answer to 10 x 10 be 100? Wait! Because our Anne Boleyn has 11 fingers, the answer 91 is in Base 11 (trust me, that's how it works). And 91 in Base 11 is 9 elevens and 1 unit, which comes to 99 + 1 = 100. The right answer!

Great news, then, because not only can you now do some tricky multiplication using however many fingers you have, but you also now have a great opening line should you ever happen to meet a 27 fingered alien at a party. While others are stumped for conversation, you can immediately step up and demonstrate how to multiply 23 x 19.

Multiplication shortcuts

What about larger multiplications? There are all sorts of short-cuts that help to simplify things. For example, most people would find the sum:

244 x 5

difficult to do in their heads. If you are one of those people, then there is a creative shortcut that you might find helpful. Instead of multiplying by 5, you get exactly the same answer if you multiply by 10 and then divide by 2. Or you can do it the other way around – first divide by 2 and then multiply by 10. This is exactly the same as multiplying by 5 because 10 divided by 2 = 5. In general it is easier to double or halve a number than it is to multiply or divide by 5, and all you are doing by changing the sum is doubling one number and halving the other one, leaving the answer unchanged.

So, for example, what if you want to do the sum 244 x 5 in your head.

Step 1: 244 divided by 2 = 122

Step 2: 122 x 10 = **1220**

What about 244 divided by 5? Again, you can turn this into a sum that you probably find easier:

Step 1: 244 x 2 = 488

Step 2: 488 divided by 10 = **48.8**

These techniques may or may not work for you. The point is that there's always more than one way of doing a calculation (contrary to what you might have been taught at school) and part of the creativity of maths is in discovering the many different ways in which you can come up with the right answer – some of which are quicker than others.

Squaring in your head

Now this isn't a textbook about doing sums, so I'm not going to start delving into how to do long division or calculate square roots. But there is one trick for doing shortcut calculations that I want to finish this chapter with because – with a little practice – you can learn it and then amaze your friends with your ability to do lightning calculations without the aid of a calculator.

First, let's look at some patterns. Which numbers add together to make 20?

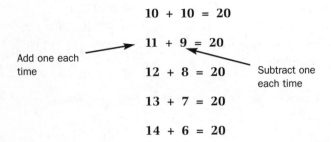

$$10 + 10 = 20$$
$$11 + 9 = 20$$
Add one each time
$$12 + 8 = 20$$
Subtract one each time
$$13 + 7 = 20$$
$$14 + 6 = 20$$

and so on. You can continue this indefinitely, even going into negative numbers if you want.

Now what happens if you multiply those pairs of numbers together? Since they add to the same number (20) maybe they multiply to the same number too. Let's see what happens to the *product* of each pair of numbers:

11 x 9 = 99

12 x 8 = 96

13 x 7 = 91

14 x 6 = 84

Hmm, as we move away from 10 x 10, the answers seem to be getting smaller. How much by?

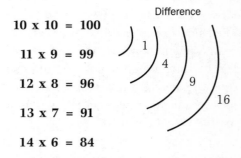

10 x 10 = 100

11 x 9 = 99

12 x 8 = 96

13 x 7 = 91

14 x 6 = 84

There seems to be a pattern: **1, 4, 9, 16**...
that's 1 x 1, 2 x 2, 3 x 3, 4 x 4, and so on.
Does this always work? What if we choose numbers that add to 24, starting with 12 and 12?

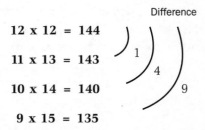

12 x 12 = 144

11 x 13 = 143

10 x 14 = 140

9 x 15 = 135

It's the same pattern, and indeed no matter what pair of numbers you start with, you will always get the same thing. If you start with a number multiplied by itself (a square, written as 12^2, for example), then as you add and subtract one from each number, the product of the two numbers reduces by the square of the difference.

And with this secret revealed, you now have the power to do some very complex squaring *in your head*.

For example, what is **99 x 99**? It's hard – at least if you do it the conventional way. But now we know a trick.

To make the sum simpler, change it from 99 x 99 to 100 x 98 by adding one and subtracting one to the two numbers. 100 x 98 is easy to do in your head, just add two zeroes to give you 9800. But hang on, by our newly discovered rule we know this number will be 1 less than 99 x 99. So the answer is **9,801**. Check it on a calculator to convince yourself.

Maybe you now have the confidence to do 98^2. Make the sum easy by going to 100, to give you 100 x 96 = 9,600. And since this time we moved 2 away from 98, we need to add 2 x 2 = 4, so the answer is 9,604. If you get more ambitious you can go as far as you like. What about 107^2? Or 89^2?

There are stage performers who make their living out of this sort of thing – doing apparently impossible calculations almost instantly in their heads. True, they do rather more impressive things, like squaring four digit numbers or finding the 13th root of a 20 digit number. But the principle is just the same – they have discovered a variety of patterns and shortcuts that make the calculations far simpler.

Calculators have made us all a bit lazy when it comes to doing sums. But worse, they have deprived us of discovering some of the beautiful patterns to be found in ordinary numbers.

Pick a card, any card

You will need: an ordinary pack of playing cards.

There are 52 playing cards in a regular pack, divided into four suits (clubs, diamonds, hearts and spades), and each suit has 13 cards, from Ace up to King.

It says a lot for human ingenuity that such a simple and portable prop is the base of literally thousands of different games that are played around the world. This vast range of games says a lot about maths, too, because one way in which the beauty of maths reveals itself is in the way that a small number of elements can be used to create such a huge and diverse range of patterns. And patterns, after all, are what card games are all about.

In fact, simple playing cards are one of the best ways of illustrating the Ah, Aha! and Haha of maths, which is why they get a chapter all to themselves.

Card games come in several categories. There are games of pure luck, such as Beggar-my-neighbour (at school we called it 'Strip Jack Naked'). There are games of skill and strategy, such

as bridge. And there's a further raft that bring in an extra dimension of psychology, of which the most popular, of course, is poker.

In every game, there's an element of finding patterns, such as collecting three of a kind or a run of ascending cards, but even more importantly there is an element of chance. Card games hinge on the unexpected. Think of the scene in *Casino Royale* when the villain Le Chiffre lays down what appears to be a winning full house, only for the crowd to gasp as James Bond lays down a straight flush. (Well, OK, like all Bond plots that wasn't unexpected at all, but you know what I mean.)

None of these surprises would be surprises if cards were arranged in a predictable pattern to start with. That is why the first thing that anyone should do before playing a game of cards is shuffle them.

You'd think that since card games have been played for hundreds of years, that the science of shuffling would be well understood, but in fact it is only in recent years that mathematicians have actually come to grips with it.

For example, give an occasional card player a pack of cards to shuffle, and they will probably do what is called an overhand shuffle, in which the cards are taken in one hand and dropped in clumps of about 10 cards at a time into the other. This will eventually mix the cards up, but amazingly, it is believed that you need to do a staggering 2,500 overhand shuffles before a pack of cards is so mixed up that there is no trace of the original sequence in which the cards were ordered.

One way to demonstrate this to yourself is to get a pack of cards and arrange it into a very obvious pattern: all the hearts ordered Ace, 2, 3, up to King, followed by the clubs, the diamonds and the spades. Now do a regular overhand shuffle a few times. If you look at the cards, you will almost certainly find clumps of consecutive cards in the same suit throughout the pack.

A more effective way of shuffling would be to take the pack of cards and fling it in the air from the top of a 10-storey building. Then go down to the ground and pick up all the cards. This so-called 52-card pick up trick (beloved of 10-year-old boys) will randomise cards better than overhand shuffling, but it is messy.

Perfect and imperfect riffle shuffles

Fortunately, there is a tidier way to shuffle. It is known as the riffle shuffle, or the Faro shuffle, and it is the kind of shuffle used by most regular card players. A riffle shuffle involves cutting the pack into two roughly equal piles, then bending the piles up with the thumbs, bringing them together, and releasing the cards carefully from both piles so that they interweave with each other.

A *perfect* riffle shuffle is one in which the pack is divided into two piles of 26 cards, with one card dropping alternately from each pile. Very few people can do perfect riffle shuffles (can you?), but those who can are able to perform some quite staggering feats.

To understand the feats of 'perfect rifflers', you need to know that there are actually two distinct types of perfect riffle shuffle.

Cut the pack into two equal piles, the 'top half' and the 'bottom half'. As you start the shuffle, you have a choice of which card to release first. If you release from the bottom-half pile first, then the card that was originally at the bottom of the pack will remain at the bottom when the shuffle is complete (and the top card will remain at the top). This is known as an 'out' shuffle, because the outer cards remain on the outside.

PERFECT 'OUT' SHUFFLE'

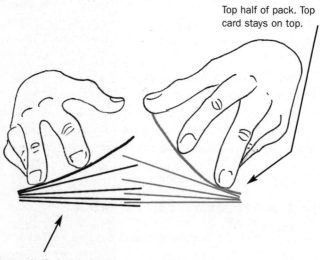

Top half of pack. Top card stays on top.

Bottom half of pack. Bottom card stays on the bottom.

On the other hand, if you release from the top pile first, the card that was originally at the top of the pack will end up second in the combined pack, and the bottom card will end second from the bottom. This is called an 'in' shuffle, because the cards that were originally on the outside of the pack are now tucked inside it.

PERFECT 'IN' SHUFFLE

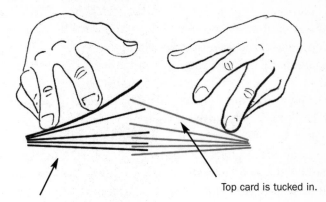

Top card is tucked in.

original bottom card is tucked in.

After a couple of perfect in or out shuffles, the cards already begin to look reasonably mixed up. But here is the remarkable thing. If you take a complete pack of 52 cards and 'out' shuffle them (perfectly) exactly eight times, you will discover you have restored the pack exactly to the order that it was in when it started! This stunt has been performed by magicians, and mathemagicians, to great effect, though it does require a huge amount of practice, because one slip and you are done for.

An easier (but still difficult) stunt is to arrange the pack into the four suits, and out shuffle it six times in preparation. Two more perfect out shuffles will turn this 'mixed' pack into a pack that is perfectly sorted into the four suits.

And if you want to make it easier still, create a smaller pack of just eight cards, arranged into the order:

Ace, 2, 3, 4, 5, 6, 7, 8.

After just three 'out' shuffles you will find the order is restored perfectly. Once you've mastered this, arrange 14 cards into a pattern, then do four 'in' shuffles, and again the cards will be restored to their original order.

It looks like magic, but in fact it's just maths.

All of this demonstrates that although riffle shuffles are good for mixing cards up, it pays for shufflers to be good, but not perfect.

Fortunately, most people are far from being perfect shufflers. A good shuffler will be able to make the cards from the two piles drop quite evenly, so that for example one card will drop from the left, followed by a couple from the right, maybe two from the left, one from the right and so on. This slightly imperfect riffle shuffling turns out to be exactly what is needed to make a pack of cards genuinely mixed up and unpredictable.

But how many riffles do you need to mix the pack up so that it is genuinely random? The answer, as it turns out, is seven. Two mathematicians, Persi Diaconis and Dave Bayer, discovered the mysterious seven when they investigated what happened to the patterns of cards when they were imperfectly shuffled.

After one, two, three shuffles, the original card pattern does begin to mix, but traces of the original order can still be spotted. Even after six shuffles, there is a chance that cards might be paired up in a predictable way – and that of course is very risky for a casino, where there are professionals who are skilled enough to follow cards and hence improve their chances of winning games. However, after the seventh decent but imperfect riffle shuffle, all vestiges of the original patterns are lost. That is why these days, casinos have adopted seven as the number of times they shuffle packs between games.

A magic riffle

So far, we've seen that eight perfect riffle shuffles can miraculously restore a pack to its original order, while seven imperfect riffle shuffles can eliminate all order and make a pack completely random.

Now I'm going to confuse things further, because there is one particular situation where, against all intuition, an imperfect riffle shuffle can leave cards with an ordered pattern that has been known to make grown adults gasp in astonishment. I recommend you to try out this little experiment straight away.

First, arrange your pack of cards so that the colour of the suits alternates between red and black. Don't worry about the values of the cards, as long as at the end you have a pack that goes red, black, red, black, red, black.

Next, cut the pack into two piles, and turn one of the piles face up. Now riffle the two piles together, so that you end up with a pack with some of the cards face up and some face down.

Now here is the mysterious thing. With the shuffled pack, pick off the top two cards. What chance would you say there is that these two cards are a red and a black? Now that you've shuffled the cards, it's reasonable to think that there could be two blacks together at the top, so the chance of one red and one black is 50–50 perhaps?

Check them, and you'll discover one red and one black. What about the next two cards? They are red and black too. In fact, as you remove cards from the entire pack in pairs, you'll discover that every single pair has one red and one black. You will never find a pair of blacks or a pair of reds, even though you riffled the pack and even though your riffle was imperfect. There WILL be times when two reds or two blacks are next to each other, but they will always belong in different pairs. For example, the top of the shuffled pack might be ordered like this:

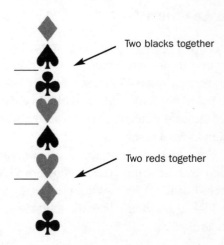

Two blacks together

Two reds together

However if you remove pairs of cards from the top, you will always end up with red and black (or black and red).

If you find that surprising, then try this. Rearrange the pack, this time putting the cards in the order: Clubs, Diamonds, Spades, Hearts, Clubs, Diamonds, Spades, Hearts... all the way through. Again, take (roughly) the top half, turn it face up, and riffle the two halves together. What chance that the top four cards are exactly one of each suit? And the next four? And the next?

You should discover something that is bordering on the miraculous. Each group of four that you take off will have exactly one of each suit, *regardless of how well or badly you riffle-shuffled the cards*. Once again, the explanation for this is mathematical. It is possible to prove that however good or bad the shuffle, .this trick will always work.

This special shuffle of a pre-arranged pack is known as a Gilbreath shuffle, after its inventor Norman Gilbreath (for some reason I get a kick from the fact that this was invented by somebody called Norman). It is the basis of a number of card tricks. It may seem to be something to do with chance and amazing coincidences, but in fact there is no chance involved at all. It is 100 per cent guaranteed to work every time, the maths says so.

An Ace miracle

There is another seemingly miraculous effect that again has nothing to do with chance. To perform this effect (and amaze yourself) you need four Aces and 12 other assorted cards.

Deal the cards onto a table *face down* like this (opposite), making sure that the four Aces are placed in the positions indicated in solid black.

Next, turn over the four cards that are in the positions indicated with shading. You should now have 12 cards face down (including the Aces) and four assorted cards face up on the table. In this little card 'miracle' the four assorted cards that are face up are going to, ahem, turn into Aces.

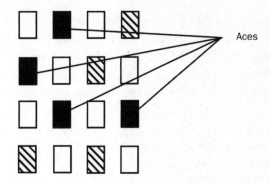

Aces

To do this, you need to fold the cards onto each other in a particular way, until you have a single pile of 16 cards. To make the first fold, choose any of the vertical or horizontal gaps between the rows and columns, and imagine folding the cards over that line. For example, you might choose the line between the second and third column of cards.

Take each card in the fourth column and 'fold it' over onto the first column, and fold each card in the third column onto the second column. (It might be easier to picture this if you imagine that the cards have been glued down onto a sheet of cling film, and you are folding the right hand columns of cards over onto the left hand columns all at once, as the diagram shows).

Now choose another gap between a row or column of cards, and again fold the cards over this invisible line. Keep doing this, folding over rows or columns, until there are just two piles left, and finish by folding one of these piles over onto the other one (which of course means you are turning that final folded pile upside down and placing it on the other pile).

If you've followed the directions correctly – and I realise that this might be feeling a bit like the instructions from an IKEA manual – you will now find you have 16 cards in a pile, 12 face down and four face up, exactly as when you started – but with one very important difference. The four cards that are face up are no longer randomly assorted cards. They are all Aces. It's as if the four cards you started with have somehow been transformed.

Before you go any further, make sure you've managed to get this trick to work, otherwise what I say next has no impact whatsoever.

I've seen this little 'card magic' work on some quite hardened adults, who are particularly surprised because they feel that they have complete control over which cards get folded over in which direction. And it's true, they do have complete control, but it doesn't matter, because whatever columns or rows they choose to fold over, the end result will be the same. They are even more surprised if, at the start, you don't draw attention to the fact that you have strategically placed four Aces in the desired positions.

Why does it work? The secret is in the relative positions of the Aces and the four cards you turned over at the start. Imagine a mini four-by-four chessboard as in the diagram opposite. If you look at the positions of the Aces and the four assorted cards in the original diagram, you'll see that they were all placed on 'black' squares of this imaginary chessboard.

If you fold the rows or columns of a chessboard onto themselves till you have a single pile of chess squares, all of the black squares will end up pointing upwards and all the whites will end up pointing downwards (or the other way round of course!). It may not be obvious why this is the case, but believe me, it's true.

If however you start by turning four of the black squares over, then those four will end up pointing the same way as the white squares, leaving only the four un-reversed black squares pointing the other way.

The four black squares that you reverse at the start are the equivalent of the four assorted cards in the trick, and the four that aren't reversed are the Aces. You might need to think about that for a while. Or have a lie down. Or just jump to the next section. Remind yourself, *there are some geniuses who don't understand this stuff either.*

Snap!, and other random surprises

In the Gilbreath shuffle, and now in this four Aces trick, surprising patterns emerged out of what seemed like chaos. However, you might feel it is a bit of a cheat, since the cards were arranged into a special order to start with.

What about a fully shuffled, truly randomised pack? Are there any patterns to be found in such a pack? The surprising answer is yes, though of course since these cards are shuffled there is an element of chance involved.

The first pattern concerns runs of red and black cards. In a fully shuffled pack, you will almost always find a run of at least five red or black cards somewhere in the pack. (Actually it should happen about three quarters of the time). If you are ever presented with a 'shuffled' pack in which the cards go red black black red black red red red black... with no colour having a run of more than three, you have every reason to believe that the pack has been rigged.

There are also patterns in 'random' games like Snap!. The whole game of Snap! is based on the fact that if you turn over two piles of cards, every so often the two cards that are turned over will be the same value (prompting the cry of Snap!). What is quite surprising is that the frequency with which you will get a snap occurring as you go through a pack of cards is predictable. If you and a friend each start with a pack of 52 cards, shuffle them, and turn them over one at a time, two times in three you will find yourselves turning over exactly the same card at the same time before you get to the end of the pack. More accurately, the chance of a snap in this version of the game is $1.718/2.718$, or 63 per cent. The number 2.718... is otherwise known as Euler's number (usually shortened to a little 'e').

Actually, most games of Snap! aren't played with two packs. The game is usually played with a single pack, and Snap! is called when two cards of the same value are turned up at the same time regardless of what suit they are. The chance of snap before the cards are exhausted in this situation can also be calculated – believe me it can and has been calculated – and it is close to 80 per cent.

A stunt to defy probability

There is one more little stunt that, when it works, will amaze even the most skeptical of audiences. In fact, it is particularly effective with groups of mathematicians. Take out a pack of cards and openly remove a particular card (without showing the face to your audience) and place it in your top pocket. If you decide to perform your stunt on a man, the card you select should be the Seven of Spades, if it is a woman choose the Seven of Hearts.

Ask your target to think of a number between one and 10, and then to name a suit. If your friend picks randomly, then the chance that he or she will name the card you picked is extremely remote, 1 in 40. But in practice, people don't think of cards entirely randomly. As much as half the time, people asked to choose a number between 1 and 10 choose the number 7, and when asked to name a suit, men typically choose Spades and women choose Hearts. For this reason there is a good chance that your friend will choose the Seven of Spades or Hearts.

If your friend happens to mention another card, you can tell them how unusual they were for not picking your card, and move onto something else. But if they happen to name the Seven you picked out, serenely remove the card from your pocket as if you knew all along what they were thinking, and say simply that there are some things that maths cannot explain. At which point they are likely to hail you as the new Derren Brown.

The back of an envelope

You will need: some A4 paper, a large envelope to put it in, scissors and some tape.

Here is a rectangle. Doesn't it look beautiful to you?

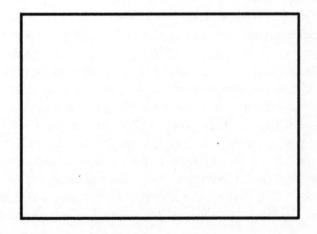

Somehow it seems to have perfect rectangle-like dimensions. In fact, if you wanted to draw the ideal rectangle, it would probably be exactly this shape, not too long and skinny, not too short and fat.

And what shape is this rectangle?

It is in fact the shape of a regular sheet of A4 paper, the sort that you stick in your printer and photocopier. It's the same shape, by the way, as its sister rectangle, the C4, which is the envelope that is just large enough to take an unfolded sheet of A4. We'll come back to C4 a bit later.

There might be some people reading this who were expecting me to reveal that this was a so-called *golden rectangle*. The golden rectangle certainly has its fans, particularly in the artistic community, but it is a different shape from the A4 rectangle and (in my entirely subjective opinion) not quite as pretty. I'll explain more about that rectangle in Chapter 11.

So what is so special about the rectangle that makes a sheet of A4 paper? Not its dimensions, that's for sure. A4 paper has the instantly forgettable dimensions of 210mm wide and 297mm long (or the equally unexciting 8.3 x 11.7 inches if you prefer to think imperial).

What is special about A4 is that when you fold it in half, you get another rectangle *which is exactly the same shape*. And if you put two A4 sheets next to each other, you get a sheet called A3, which is also the same shape. This is the only shape of rectangle that has this elegant property, and its secret can be found if you divide the length of the sheet (297) by its width (210) – it gives you the number 1.4143. And if you multiply 1.4143 by itself you get? 2.00024, which is suspiciously close to 2. This is no coincidence, for the ratio of the sides of A4 paper, or any other sheets in this series, is always the square root of 2.

Root-two is, to use the mathematical name, an irrational number, because it cannot be represented as a fraction of two whole numbers. (297/210 comes close, but is not quite right).

It may be irrational, but the use of root-two in A4 paper is extremely rational. That's because if you keep doubling up A4 sheets, you create A3 paper, then A2, and A1 and you finally get to A0 paper, which has an area of exactly one square metre. In other words, an A4 sheet has an area of exactly one sixteenth of a square metre. You never know when that snippet of knowledge might come in handy.

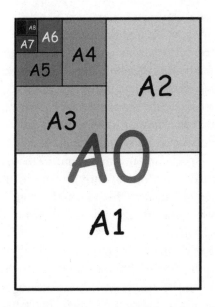

If you start folding sheets of A4 in half, the series continues in similar logical fashion to A5, A6 and so on, for as long as you wish to go. An A10 sheet is about the same size as a postage stamp. Another couple of folds and you get A12, so tiny you'd probably lose it. This does, however, raise the age-old question: how many times can you fold a piece of paper in half before it becomes impossible to fold it any more?

Multiple folding

The common belief is that the maximum possible number of times you can fold any piece of paper is seven or eight. And for many years, nobody thought to challenge this belief. Then in 2002, Britney Gallivan, a high-school student in California, discovered a way of folding a piece of paper not eight times but a whapping 12 times, having first worked out a mathematical formula that showed how many folds were theoretically possible.

It's worth pointing out that 12 folds of an A4 sheet are not possible, but if you use a very long piece of toilet paper (hundreds of metres long, in fact) as Britney did, then you are in business. Incidentally, folding over and over in the same direction turns out to enable you to do more folds than alternately folding along and across.

Paper-folding is usually thought of as an art, but as Britney Gallivan showed, it is clearly a science too.

In fact it can be an art and a science at the same time, as I'm now going to demonstrate. You are about to discover that it is possible to produce something of stunning mathematical beauty by simply folding a long strip of paper in half lots of times. And what we are about to create also has a surprising connection to the novel *Jurassic Park*. I hope that has whetted your appetite.

At this point you might like to get a C4 envelope and tear a 2.5cm wide strip from it that 'wraps around' the bottom of the envelope to making the strip twice the length of the envelope. For what you are about to do, the longer the strip of paper the better.

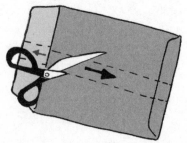

If your strip doesn't already have a crease down the middle, fold it lengthways, right over left, like this:

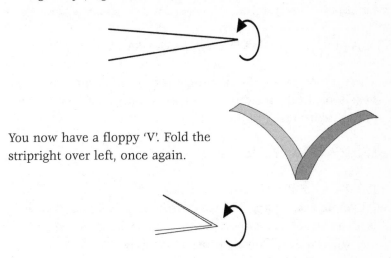

You now have a floppy 'V'. Fold the stripright over left, once again.

Keep repeating this right over left fold until you can fold no further. At each stage, unfold the paper to see what pattern it is making. You will find that the creases make a series of valleys and peaks. If you do it correctly, this is the series of patterns that will unfold:

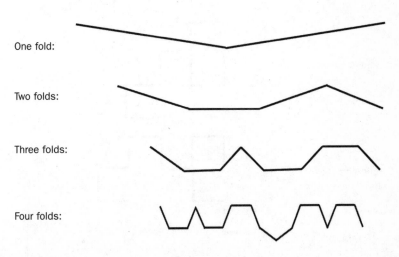

One fold:

Two folds:

Three folds:

Four folds:

Can you tell where this is going? I hope not! There is actually a pattern, which you might be able to spot if you call the upward kinks (or valleys) 'V' and the peaks 'Λ'. The sequence goes:

V, VVΛ, VVΛVVΛΛ, VVΛVVΛΛVVVΛΛVΛΛ.

The rule for getting from one member of the sequence to the next is to take the pattern, rotate it half a turn, stick it on the end, and then insert a V in the middle.

For example: **VVΛ** becomes **VVΛ V VΛΛ**

But that's not the interesting part.

   Where this folding of a paper strip gets really mind-blowing is if you bend the creases so that they all make right-angles. Now you will see the following pattern emerging:

First iteration:

Second iteration:

Third iteration:

Fourth iteration:

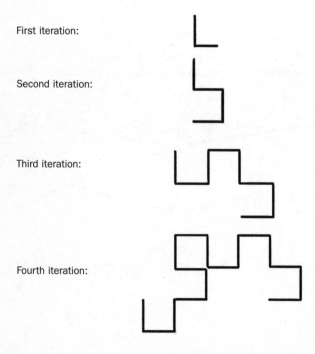

By the time you get to the seventh iteration, it looks like this:

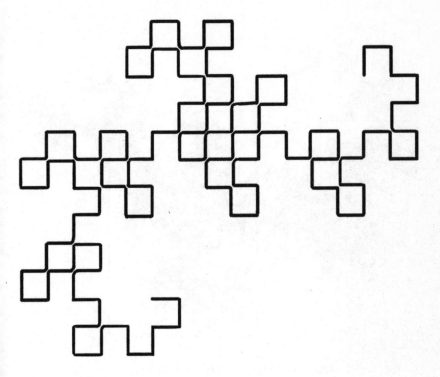

Suddenly this is becoming a complex shape. And look at that block of six squares in the middle of the pattern. What seems rather eerie is that the paper never crosses over itself: you could carefully pull the two ends, that are still visible, and the whole thing would unravel to the original strip without any tangles. It is tempting to ask how the paper 'knows' how to bunch itself into a grid of squares, without ever getting tangled.

The seventh iteration is, of course, about as far as most paper folding can get before the strip of paper becomes too stiff to fold any more. But what if we take one of Britney Gallivan's long pieces of tissue and fold it a record-breaking 13 times. What pattern emerges? The answer is little short of stunning, for the pattern after 13 iterations is this:

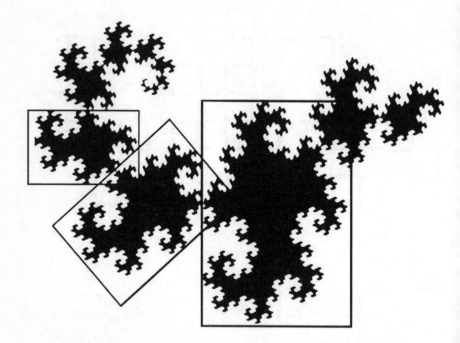

The folds have become so tiny they are now almost invisible, but a beautiful shape has emerged. What you see here is a pattern known as a fractal, in which the same shape repeats itself in smaller and smaller versions.

The more folds that you do, the more delicate this fractal pattern becomes, until ultimately the jagged edges become beautiful and intricate curves. And there is another curious feature: if you draw rectangles around each replicated shape, the edges of each rectangle get smaller by a factor of 1.4143 each time. That's the square root of two again, the same number that featured in our original piece of A4 paper. That's a coincidence, though maths does have a habit of revealing certain numbers again and again in different situations.

The fractal pattern that comes from simple paper folding is known as a 'dragon curve' – because its discoverers William Harter and John Heighway thought it looked a little like a dragon.

The dragon curve also features in the book *Jurassic Park* (though not in the film of the same name). If you dig out a copy of Michael Crichton's novel, you will find that the book is divided into sections, and each section has an illustration at the start. These illustrations are iterations of the dragon curve, identical to the iterations that featured in earlier pages – though for a reason that he doesn't explain, what I have called the fourth iteration, Michael Crichton calls the first iteration.

But why does this pattern feature in *Jurassic Park*? The reason is that it has a metaphorical link to the story. In the novel, the scientists start with something very simple, dinosaur DNA, and use it to generate more and more dinosaurs.

What they fail to realise is that sometimes a simple procedure can have a quite dramatic and unexpected consequence. In the case of *Jurassic Park*, the monsters begin to mate, and the community of dinosaurs takes on a frightening pattern all of its own. In the case of folding a piece of paper, the result is a (rather less frightening) sort of monster.

Twisting instead of folding

OK, what else can we do with a strip of paper torn off the back of an envelope? Instead of folding it, you can twist it. If you take a strip of paper, give it a half twist, and tape the ends together, you get what is called a Mobius Strip, or an Afghan Band. It's quite likely you first came across this when you were about eight years old when the teacher got you to make the strip, and then to carefully cut along the middle of the strip all the way around. You might expect that this would create two paper hoops, but instead when you cut a Mobius strip in half you get one big loop of paper. It's quite a wow, at least it is the first time you do it. (And I know a seven-year-old who was still being wowed by it the 20th time he did it, having devoted an entire afternoon to making Mobius Strips and cutting them in two. I think he half expected to 'catch it out' one time, but he never did.)

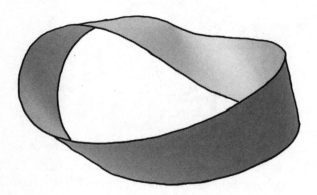

If you find the Mobius strip a bit hackneyed, then you will be pleased to hear that there is a variation that has a nice kick to it. I call it the Motorway Strip. You need a long-ish strip of paper (a double length torn carefully from a C4 envelope works nicely), and you should make it fairly wide, say 10cm.

Before making the twisted strip, draw dotted lines along one side of the strip, as if it were a three-lane motorway. Now do a half twist and tape the ends together to create a Mobius Strip as before.

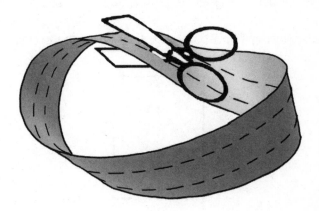

This time instead of cutting along the centre of the strip, you cut along the dotted line of the 'slow lane' of the motorway (as it were). Keep on cutting, and after a while you will discover that you have switched over to the dotted line of the fast line. Eventually you'll get back to where you started. What do you get? Intuition (our old, unreliable friend) might say you are going to get three rings, or perhaps one giant ring. But in fact you get neither. Instead, you get one large hoop that is linked to one small hoop half its size. It looks a bit like an Olympic medal, especially if you hang the large loop around your neck.

Indeed, if you make the Motorway Strip using a striped school scarf – ideally one with two bands of black on the outside and a band of gold down the middle – then when you make the twisted band and cut it as before, you'll discover that the small hoop comes from the central gold band of material.

The crisp bag surprise

I have one final paper folding surprise. This time you can use the entire envelope (though you will be snipping a bit of it off). You can turn any envelope, not just a C4 envelope, into a very special type of pyramid. To do it, follow these instructions:

Fold the envelope in two to make a crease down the centre.

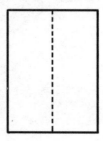

Take the bottom left corner and fold it up so that it touches the dotted centre line, and makes a straight line to the bottom right corner. Mark the point where the corner touches the centre line with an X:

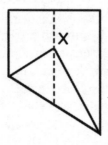

Fold the envelope from the top horizontally along the line through X and make a firm crease. Then cut off the rectangle above this crease and throw that part away.

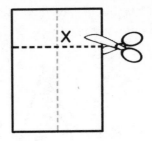

Don't lose faith, we're getting there! Now take the top left corner of the remaining portion of the envelope and fold it so that it makes a diagonal edge between X and the bottom left corner, and firmly crease this.

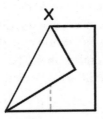

Repeat this for the top right corner of the rectangle, making a diagonal edge from X to the bottom right corner.

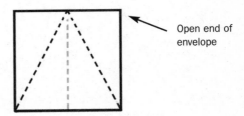

Open end of envelope

You should now have an envelope with creases that form a triangle. Think of it as a crisp packet that you want to open up – do this by pressing in the two edges to open the envelope's 'mouth'. If you keep pressing, the two edges will join up, and you should tape the two sides together.

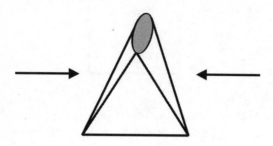

And after all that, what have you got? No rectangles at all – all the rectangles of the envelope have disappeared to leave you with nothing but triangles, forming a perfect pyramid for which each face is an equilateral triangle. Not the intricate beauty of the dragon curve, but a different, simple beauty all of its own.

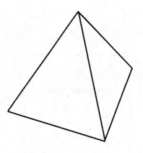

5

What colour was the bear?

You might need (but probably won't): a clock, a globe and a piece of string.

What's the difference between a puzzle and a maths question? The truth is that sometimes it's hard to tell, though most people would reckon they could spot the difference. Typically a maths question is something that is done as a chore, whereas puzzles are done for fun. But for me, the mark of a great puzzle is the

same as the mark of a great maths question, that it contains an element of surprise, beauty or humour – some ah, aha! or haha. Here are six puzzles. In each case, there doesn't appear to be enough information to answer the question (but there is!). In a couple of them, the question doesn't even appear to have anything to do with maths.

The bear

A man leaves his tent. He walks 1 mile south. Then he walks 1 mile east. Then he walks 1 mile north.

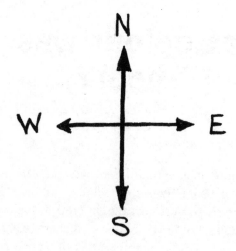

To his surprise, he now discovers that he is back as his tent, and at this moment he spots a bear. What colour is the bear?

Meanwhile, another man (camping elsewhere) gets out of his tent, points north and walks 1 mile in a straight line. Then he sits down for a cup of tea. When he's finished the tea, he points north again and walks another mile in a straight line. To his amazement he discovers he too is back at his tent. Any chance of a bear this time?

What's the time Mr Wolf?

Mr Wolf has a clock, but he has removed all of the numbers. Not only that, but he has turned the clock around, so that the 12 is no longer at the top – but he refuses to say where the 12 now is. Here is the clock. What's the time Mr Wolf?

The Monk

There is a narrow winding path that leads up a mountain to a remote monastery. One Monday morning, at sunrise, a monk sets off up the mountain. Every so often he stops to admire the view, or to pick a flower, or to have a drink and a bite to eat. Finally, after meandering up the mountain all day he arrives at sunset at the monastery, where he spends the night. On Tuesday morning at sunrise he sets off down the mountain. Walking downhill is faster, but he does take an extra long lunch break and he spends a couple of hours sitting on the path reading a book. At one point he even retraces his steps up the path for a few minutes to find a button that has dropped off his tunic. Finally at sunset he gets to the bottom of the path.

Over the two journeys, the monk walked at different speeds and all we know is that each day he started and finished his journey at the same time. The question is, was there ever a time of day on the Monday and the Tuesday when he was in exactly the same point on the path?

The burning string

You have a piece of string, and you have been told that if you set fire to one end, it will take exactly one hour to burn to the other end. You need to be able to measure exactly 30 minutes and don't have a watch, just the string and a box of matches. Easy, you might think, just fold the string in half and measure how long it takes for half of the string to burn.

Unfortunately there is a catch: the string is very uneven, in some places it is thin and in other sections thick, so the rate at which it burns varies. For all you know, the first half of the string will take only five minutes to burn and the second half will take 55 minutes. On the face of it, it seems impossible to find a way of measuring exactly 30 minutes, but there turns out to be an ingenious solution. Can you find it?

The Saddam hat puzzle

A cruel dictator – let's call him Saddam – has 100 prisoners. One evening he gathers the prisoners together to announce to them that the following day he plans to execute some of them.

'I will line you up and then bury each of you up to your neck in the sand, each of you facing forwards, so that the prisoner at the back will be able to see the 99 other prisoners, but the one at the front will not see anyone. Next I will come behind each of you and place on your head a hat that will be either white or black. You won't be able to see the colour of your own hat, and I have a vast supply of black and white hats, so it is possible that each of you could have a white hat, or each of you could be black, or it might be a complete mixture.

'You must remain silent at all times. Once you all have your hats, I will start at the back of the line, and ask each prisoner to call out a colour, either Black or White. If you correctly call out the colour of the hat on your head, you live. If you get it wrong – BANG!! – you die.'

Saddam then leaves. The prisoners have the evening to try to come up with a plan to save as many of themselves as possible.

Soon they think of a plan. The person at the back will call out the colour of the hat in front of him, which means that prisoner in front will know his colour. Then the second prisoner will call out the colour in front of him, and so on all the way to the front. Then they realise this cunning plan has a major flaw. If they call out the colour in front and it is different from their own hat's colour, they might all get shot. So they are stumped.

Can you think of a better plan? How many prisoners can be guaranteed to survive out of the 100 prisoners? (Remarkably, it is more than 50.)

Where is the father?

Finally, here is a puzzle that tends to get a particularly high 'haha' score. A woman is 21 years older than her son. In six years time, she will be five times as old as her son. Where is the father?

ANSWERS

The bear

The bear is white.

Normally, if you travel 1 mile south, then east then north, you
don't end up where you started – you end up roughly 1 mile east
of where you started. The only place where this is not the case
is near the North (or South) Pole. If you start on the North Pole,
then travel 1 mile south, 1 mile east and 1 mile north, then you
end up at the North Pole again. The only bears to be found near
either of the poles are polar bears, hence the bear's colour.

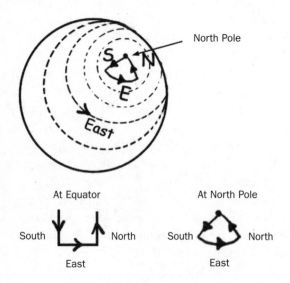

There could be a polar bear for the second man, too.

The only place where the second stroll can take place is any-
where within a mile of the North Pole except for the pole itself.
When the man points north and walks for a mile he goes over
the pole and keeps on until the mile is done. When he points
north again he is simply retracing his steps towards the pole
and back to his tent.

Sadly, these puzzles could become an unlikely victim of global warming, since the disappearance of the polar ice caps will make it impossible for bears to get to the North Pole, let alone for a man to pitch his tent there.

What's the time Mr Wolf?

It's Dinner Time! (as they say in the popular children's game.). The clock appears to say quarter-to-four, but we are told that it has been rotated, in which case what is to say that it isn't twenty-to-three, or something o'clock? In fact there is one very good reason why it cannot be any of these times, if you look at the hour hand, which is pointing midway between two hour marks. That will only happen when the time is half past the hour. Rotate the clock around so that the minute hand points straight down, and the time is revealed to be 12.30, which is lunchtime – otherwise known to some as 'Dinner Time'.

There are many puzzles relating to the hands of a clock dating back to Victorian times. One of the best, because it sounds so simple, is the question: how many times does the minute hand overtake the hour hand between midnight and noon? Since the hour hand goes past 11 numbers in this interval, the natural answer would appear to be 11, but the correct answer is 10. The first occurrence is just after 1.05, the next is around 2.12, then 3.17, each time the hour hand moving further and further around from the hour that it last passed. The 10th time that the hands cross is at just before 10.55, and the next time they meet is noon – but the minute hand doesn't overtake at this point, it is level.

The Monk

Yes, there is certainly a point on the path where the monk passes at the same time on both days.

There doesn't appear to be enough information – after all, we don't know what speed he walked up and down the mountain, nor what times he stopped for his different breaks. But none of that matters, if you imagine superimposing the two days on top of each other. Imagine making a film of the monk doing his Monday climb. On the Tuesday, we now project an image of the Monday monk walking up the mountain, while the real monk begins his journey down the mountain, both of them setting off at sunrise. Since both monks eventually reach their destination at sunset, *there must be some moment when the two of them pass each other on the path*. It doesn't matter that we don't know what time this happens, we can still be confident that it does happen at some point.

The Burning String

Set fire to both ends of the string. When the two flames meet, 30 min-utes must have elapsed.

String (or 'shoelace') puzzles like this took off in the 1990s, their popularity guaranteed because of the 'aha' moment of insight that is required to solve them. A slightly harder variant of the puzzle requires you to use two identical 'one hour' strings to measure exactly 45 minutes. In this case the solution requires an extra creative step. At the start there are four ends on the two strings, and you should light three of them, two on String A and one on String B.

When the flames on String A meet, 30 minutes have passed.

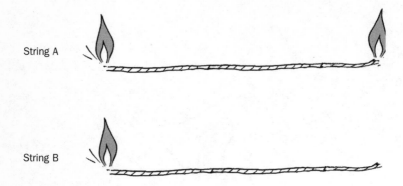

String A

String B

At this point, light the other end of String B, and when the two flames on String B meet, half of the remaining time (15 minutes) has passed, making 45 minutes from the time the first ends were lit.

Saddam Puzzle

Whatever hat pattern Saddam chooses, 99 of the 100 prisoners can be saved, and even the 100th has a 50–50 chance of surviving. The method of achieving this remarkable survival rate is ingenious – I have never met anyone who worked it out before seeing the solution. To make the method work, the prisoners need to accurately follow three rules.

Rule 1. Each prisoner looks at the line of prisoners in front of him, and counts the number of white hats that he can see. If he can see an EVEN number of white hats then he calls his own hat BLACK. If he can see an ODD number then he shouts WHITE. (The prisoner at the front can see zero hats, an even number, so he calls his hat BLACK.)

Rule 2. Starting at the back, each prisoner must shout out the colour he currently thinks he is, loud enough that the prisoner at the front of the line can hear him.

Rule 3. Every time that a prisoner hears one of the prisoners behind him shout WHITE, he switches the colour he is currently thinking of. (So the first time a prisoner labelled Black hears WHITE shouted out, he switches to WHITE, the second time he switches to BLACK and so on.)

And that is it – the method is entirely self working. To demonstrate it, let's imagine a line of just six prisoners, like this:

Each of the prisoners starts by counting how many white hats he can see in front of him, using Rule 1. The prisoner at the back (on the left in the picture) can see an even number of white hats (2) so he calls himself BLACK. The second prisoner can see an odd number and calls himself WHITE, and so on. Before any prisoner has shouted out, they think their colours are as follows.

Now they begin to announce their colours. The prisoner at the left shouts out BLACK. Sadly, he is done for. The second now shouts WHITE and is saved – and on hearing the WHITE shouted, each of the other prisoners switches colour (as in Rule 3):

B   W   B   B   W   W

These four prisoners all switch
colour when they hear Number 2
say White

Next, prisoners 3 and 4 correctly call BLACK and number 5 calls WHITE, at which point number 6 switches colour to become BLACK. Apart from the prisoner at the back, who was unlucky that the colour he called was different from what he was wearing, each of the others is saved. This is the case however many prisoners there are in the line.

If you do ever find yourself in the unfortunate position of being a prisoner in this situation, there are a couple of things to bear in mind. First of all, don't volunteer to be the prisoner at the back. In fact, you want the person who does volunteer for that role to have a number of special qualities: he must be prepared to die for the cause, and he must have good enough eyesight that he can see all the hats up to the front. He has to be able to count, too.

As long as you can trust the prisoner at the back, then remarkably your own position is now secure. What if one of the prisoners behind you (apart from the back one) makes a mistake and gets shot? No problem – if you hear the bang of the gun, just treat it as if it is another prisoner shouting WHITE and switch the colour of your hat. When it comes to your turn, you will call out the correct colour of your hat.

Like a number of ingenious puzzles, this one almost certainly originated in the world of computing. The method used to save the prisoners is similar to techniques used to ensure that streams of data (computer data is always a string of ones and zeroes) end up correct, even if some of the data is corrupted.

Where is the father?

6. *The father is...very close to the mother*. At first glance it seems ridiculous that you could possibly know where the father is, but work through the numbers and all will be revealed.

Call the age of the Mother M and the age of the son S. The puzzle says that: The mother is 21 years older than the son, i.e.

$$M = S + 21$$

In six years, she will be five times his age, i.e.

$$M + 6 = (S + 6) \times 5$$

Now for some algebraic manipulation – bear with me. Replacing M with S + 21 and multiplying out the brackets gives you:

$$S + 27 = 5 \times S + 30$$

And subtracting S + 27 from both sides leads to:

$$0 = 4 \times S + 3$$

Finally, subtract 3 from both sides, to give:

$$4 \times S = -3, \text{ or } S = -\tfrac{3}{4} \text{ (a negative number).}$$

In other words the Son is $-\tfrac{3}{4}$ of a year old. What does this mean? $-\tfrac{3}{4}$ of a year is minus 9 months, in other words the son is still nine months away from being born.

The current location of the father is left to your imagination.

Heads or tails?

You will need: several pennies.

When it comes to making a decision, most of us readily resort to flipping a coin. Folk have been flipping coins since Roman times. In those days, the call wasn't 'Heads or Tails' but *'capita aut navia'* meaning 'Heads or Ship' (early Roman coins had two heads of the god Janus on one side and a single ship on the other). So we probably owe it to the Romans that we still refer to Heads or Tails in the plural rather than Head or Tail, which (if you think about it) would make more sense.

The great thing about tossing a coin is that it is fair to both parties, the chance of it falling as Heads and Tails both being 50 per cent – otherwise known as 50–50, 1 in 2, a half, 0.5 or (if you are a bookmaker) 'Evens'. That's why tossing a coin is also one of the easiest ways of getting introduced to the fascinating, complex world of probability and its language.

So what surprises could our humble 50–50 heads or tails coin possibly have for us? As it happens, plenty.

For a start this popular notion that coins have a 50–50 chance

of coming up heads or tails turns out not to be true. First, you have to allow for the fact that there is a chance – albeit a very small chance – that the coin will end up on its edge. This is what happens in the film *Mr Smith Goes To Washington* when the Governor tosses a coin to decide between two candidates; when it lands on its edge he chooses a third candidate (Mr Smith, played by James Stewart) instead.

A PSYCHOLOGICAL TIP
(Piet Hein, mathematician, scientist and writer)

Whenever you're called on to make up your mind,
and you're hampered by not having any,
the best way to solve the dilemma,
you'll find, is simply by spinning a penny.
No – not so that chance shall decide the affair
while you're passively standing there moping;
but the moment the penny is up in the air,
you suddenly know what you're hoping.

A Psychological Tip by Piet Hein published with kind permission of Piet Hein A/S

Even if we discount the freak chance of the coin landing on its edge, it has been discovered that if you toss a coin in the conventional way, by flipping it with your thumb, there is roughly a 51 per cent chance that it will land with the same face showing as it started with. This is because in some flips, the coin doesn't actually rotate, it merely rises and falls like a wobbling flying saucer. So next time you are asked to call, check which face is pointing up in the flipper's hand and call that one – you will marginally improve your chances, and every little helps. (Note – if the coin flipper's technique is to catch the coin and turn it over, you should call the opposite).

What happens if you try to remove this human error by spinning the coin on its axis instead, using a smooth table?

It turns out that this is even more biased. Depending on the currency you use, the results can be a long way from 50–50. If you spin an American penny, it has been found to end up showing tails as much as 80 per cent of the time. British pennies are also heavily biased towards tails.

Another rather eccentric way of performing Heads or Tails is to rest a penny on its edge on a flat table, and then shove the table from the side causing the penny to fall over. (It gives a whole new meaning to the old game 'shove ha'penny'.) This time, it is a head that turns up the majority of the time.

So if you can get yourself a suitably gullible opponent, you can ask them to nominate Heads or Tails, and depending on which they choose, you can then decide either to spin or shove the coin to greatly improve your chance of winning.

The reason why ordinary coins have an element of bias in them is because there is typically a little more metal protruding on one side of the coin than the other. On the American penny, the blob of copper/zinc that makes up the head of Abraham Lincoln is larger – and therefore heavier – than the blob used to make the image of the Lincoln Memorial on the Tail side. Although I've not tested them, it would be surprising if Roman coins weren't just as biased themselves.

Given this tendency for coins to be slightly biased, what can be done to make the toss of a coin perfectly fair to both parties? It turns out that there is an ingenious technique that was suggested to me by my friend John Haigh. Let's suppose you have an extremely biased coin, that turns up Heads, say, 80 per cent of the time and Tails 20 per cent of the time.

To remove the bias, you need to be aware that regardless of how biased a coin is, the chance of getting a Head followed by a Tail will always be the same as the chance of a Tail followed by a Head.

All you have to do, therefore, is adapt the game of Heads or Tails to 'Head/Tail or Tail/Head'. Call out one of these, then get your colleague to toss the biased coin twice, and you win or lose with a 50 per cent chance depending on whether the coin falls Head/Tail or Tail/Head in the two flips. If it comes up Heads/Heads or Tails/Tails, you simply scrap these throws and flip the coin twice more.

Even if coins are random, people are not

Suppose now that you have a perfectly fair coin, with no bias, no chance that it will land on its edge, and a perfect way of flipping it. This coin can now be called 'random', which means that whatever has happened in the past, the chance of the next toss of the coin being Heads is exactly 50 per cent. This seems perfectly simple and obvious, yet in reality human beings have great difficulty believing that it is true. Take the following sequence of four flips of a coin:

Tails – Tails – Tails – Tails...

What do you expect to come next? It will be Heads or Tails, equally likely of course, yet an inbuilt intuition drives most people to think that a Head is somehow 'due'. Another Tail would provoke a gasp of amazement, and a sense that some greater force is at work.

You can test the inability of people to believe in true randomness of coins with an experiment. You need to ask two people to pretend to be coins. Get the two to take it in turns to call out a random coin flip (Head or Tail), and get them to keep doing this for 32 calls.

This is a typical example of a sequence that two adults (or older children) will produce:

**H T T H T H T H H H T T H T H T H T T H T H H H T T H T
T H T T H H**

Looks perfectly random, doesn't it? And this sequence of 32 coin flips is just as likely as any other sequence, so in that respect there is nothing wrong with it. But it does have a feature that makes it unusual for sequences of 32 flips. Examine it more closely, and look for runs where Heads or Tails appear more than twice in a row. There are two places where Heads turn up three times, none where Tails appear three times. And there is not a single instance where either Heads or Tails appear four or more times in a row.

This is very unusual. The maths is difficult, but it turns out if you flip a coin 32 times, then you should expect there to be a run of at least four Heads or four Tails somewhere in the sequence. Only rarely, perhaps one time in 10, will you end up with a sequence in which the longest run is smaller than four. You can even use this curious fact to test sequences of Heads and Tails for whether they were produced by a real coin or a human being pretending to be a coin. It will only work if the person doesn't know about the four-rule, but so long as that is the case, you'll find your detective work will succeed most of the time.

Which is most common? A *common* point of confusion

Every sequence of coin tosses is equally likely – so for example you are just as likely to get H H H H as you are to get T H T T. However, this often leads to a point of confusion. Although every sequence is as likely as any other, if you count up the total number of Heads and Tails in a particular sequence, some combinations are more likely than others. For example, if you flip a coin twice, there are four possible, and equally likely, outcomes:

H H
H T
T H
T T

However, while there is only one way of getting two Heads or two Tails, there are two ways of getting one Head and one Tail. To summarise, if somebody asks which is more likely, to flip two Heads in a row or to flip a Head then a Tail, the answer is *neither – both outcomes have the same chance*. But if they throw two coins in the air and ask which is more likely, that the two coins both end up as Heads, or that one ends up Heads and the other Tails, then the answer is that 'one Head and one Tail' is twice as likely as two Heads.

It takes a clear head, and perhaps a deep breath, to understand the subtle difference between the two questions, which is one reason why many people find the whole area of probability so difficult.

Penney Ante

If you are already beginning to feel your head spin (no pun intended) then perhaps you should skip this next section, because it has been known to baffle even the experts. Nevertheless, I'm about to show you one of the most remarkable curiosities relating to coin flipping. It concerns a game that is known as Penney Ante. Although it can be played by flipping a penny, that's not where its name comes from. The game was first analysed by a mathematician called Walter Penney back in the 1960s.

Penney Ante is a game for two players – you, and somebody that you wish to beat! The idea is simple enough. If you flip a coin three times, there are eight possible and equally likely outcomes:

H H H
H H T
H T H
T H H
H T T
T H T
T T H
T T T

To start the game, you ask your friend to pick any of the above sequences of coins. You then choose a different sequence. You now flip a coin, and keep flipping it until either your or your friend's sequence appears. The winner is the player whose sequence appears first.

For example, suppose your friend picks H H T and you pick T H H. You now flip the coin and the results are like this:

H T T H T H H

The game stops here, and you win, because your sequence of T H H has just appeared, and there is no sign yet of your opponent's H H T.

Now we have already established that a sequence of T H H is just as likely to appear as the sequence H H T. So it seems as if this is a perfectly fair game, with each of you having a 50–50 chance of winning. However, it turns out that there is a strategy that will enable you to win well over half the time *regardless of what your friend chooses*. The secret is that you need your friend to choose first. This might seem like you are being generous, but in fact it is what gives you the advantage, because you choose your sequence depending on what your friend chooses.

Suppose your friend goes for T T T. If this is his choice, then you should select H T T – and if you do, then you should expect to win this game seven times out of eight.

You probably won't believe me, so I suggest you immediately test it out with a coin. Play the game a few times, and I'm prepared to bet that you will find that the sequence H T T comes before the sequence T T T in almost all of your games.

But *why*? The easiest way to understand it is to imagine you are playing a game with your friend, and you've already tossed the coin a few times, when finally the sequence goes .....T T T. Hooray, it looks like your friend wins. But hang on, if this is the first time that T T T has appeared, what came immediately before it? It can't have been a Tail, because that would mean your friend won the game one flip earlier, and we've just said that this is the first time T T T has appeared. So the previous flip must have been a Head, which means that just before T T T appeared, there was a H T T. In other words *you win*. In fact, the *only way* that your friend can win with T T T is if those are the first three flips of the coin. Otherwise, H T T is bound to come earlier.

The chance of flipping three Tails in your first three throws is 1 in 8, about 12 per cent, so that is the chance of your opponent winning. That leaves an 88 per cent chance that you will win.

It turns out that for every choice your friend makes, there is a sequence you can choose that gives you the edge. The sequence you should select, and your chance of winning the game, is given in the table below:

How to win at Penney Ante

Friend chooses	You then choose	Your chance of winning
T T T	H T T	$\frac{7}{8}$
T T H	H T T	$\frac{3}{4}$
T H T	T T H	$\frac{2}{3}$
T H H	T T H	$\frac{2}{3}$
H H H	T H H	$\frac{7}{8}$
H H T	T H H	$\frac{3}{4}$
H T H	H H T	$\frac{2}{3}$
H T T	H H T	$\frac{2}{3}$

This is certainly one of those mathematical phenomena that is hard to explain in simple terms, but the great thing is that regardless of whether you understand it or not, it works! A few games played using the above strategy and you should end up quids in. Though it isn't guaranteed – that's the great thing about chance, even if you use the best possible strategy, there is always the possibility that you will be unlucky and lose.

Tossing lots of coins – and Pascal's Triangle

So far we've looked at what can happen when you toss one coin, two coins and three coins. What if you increase the number of coins that are tossed? The number of different possible sequences of Heads and Tails increases with each extra coin. In fact it doubles each time. As you've already seen, with one coin there are two possible outcomes (Head or Tail), with two coins there are four, with three coins eight and so on.

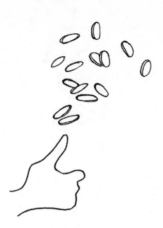

One coin

Head(1) Tail(1)
2 outcomes

Two coins

H H(1) H T(2) T T(1)
4 outcomes

Three coins

H H H(1) H H T(3) H T T(3) T T T(1)
8 outcomes

Four coins

H H H H(1) H H H T(4) H H T T(6) H T T T(4) T T T T(1)
16 outcomes

The figures in brackets show the number of ways that you can achieve that particular combination of Heads and Tails. For example, there are four ways of getting H H H T, which are: H H H T, H H T H, H T H H and T H H H.

If you write out those bracketed numbers line by line, they form a triangular pattern. This pattern of numbers is known as Pascal's Triangle.

0 coins				**1**[2]				
1 coin			**1**		**1**			
2 coins		**1**		**2**		**1**		
3 coins	**1**		**3**		**3**		**1**	
4 coins	**1**	**4**		**6**		**4**		**1**

It's easy to write out each line of Pascal's Triangle once you know the secret. Each number in Pascal's Triangle can be found by adding together the two numbers in the line above it. So you could use it to find out the possible combinations of Heads and Tails in five, six or more coins, if you were so inclined.

This symmetrical pattern of numbers doesn't just apply to coins. In maths, patterns have a habit of cropping up again and again in apparently unrelated places, and Pascal's Triangle is no exception. You will be meeting it again soon.

Notice how each line in the triangle is a palindrome, i.e. it reads the same forwards and backwards. That's a topic that deserves a chapter all to itself...

---

[2]If you toss zero coins there is only one possible outcome, which is that you get zero Heads and zero Tails!

Palindromes and other pretty patterns

You will need: a pocket calculator.

Sometime in the first century AD, a Roman graffiti artist scrawled a square grid onto a wall in the doomed town of Pompeii, and filled it with letters. It read as follows:

S A T O R
A R E P O
T E N E T
O P E R A
R O T A S

To this day, nobody knows why he wrote it, nor what it is supposed to mean, though roughly translated it says: 'The sower Arepo holds the wheels with effort'. The word square has a rather nice feature that the words can be read horizontally or vertically, and if put together in sequence they form a palindrome, a sentence that reads the same forwards and backwards:

SATOR AREPO TENET OPERA ROTAS

There are some who believe that this square was intended as a cryptic Christian symbol, because all the words combined make an anagram of PATERNOSTER ('Our father' in Latin, the first two word's of the Lord's Prayer) and A O (twice) with A O taken to mean 'from Alpha to Omega'. Unfortunately, while this early Christian connection would appeal to those who enjoy the cut and thrust world of conspiracy theories, it is probably no more than a coincidence (after all, Christianity didn't become a feature of Roman life until long after Pompeii had been buried by volcanic ash).

Far more likely, the Pompeii graffiti artist was indulging in a little word game, because the Romans are known to have loved palindromes, as did the Greeks before them.

It is sometimes joked that this ancient pastime goes back to the Garden of Eden, when the man approached the woman and said:

'MADAM I'M ADAM'

And the second palindrome came immediately as she looked him in the eye and replied:

'EVE'

And it has been going on ever since. Sometime around 1881 – a palindromic year – somebody had an idea for a great piece of civil engineering that would join the Atlantic Ocean to the Pacific through the middle of the Americas. That story is summed up nicely like this:

A MAN, A PLAN, A CANAL, PANAMA!

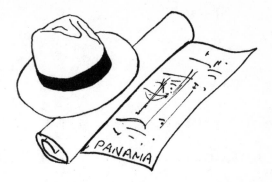

And palindromes are still being invented, my favourite being *Go Hang A Salami, I'm a Lasagna Hog!* which became the title of a delightful cartoon book by Jon Agee.

But enough of this frivolity! I just wanted to make the point that palindromes are a universal source of fun, partly because they involve delightful yet pointless patterns.

And exactly the same palindromic pleasure can be found in numbers.

Grab a calculator. Now do the following sum:

$$3 \times 7 \times 11 \times 13 \times 37.$$

Isn't that nice? Patterns that emerge from nowhere can be both mystifying and beautiful. Sometimes those patterns have meaning and sometimes they don't. To a mathematician, this difference is extremely important, but to the non-mathematician, a pretty pattern is striking whether it's there for a reason or not.

The number 11 featured in that numerical example is of course a palindrome, and it is the root of a number of other appealing patterns, too.

11 also features in a striking triangular pattern.

$$1 \times 1 = \qquad\qquad 1$$

$$11 \times 11 = \qquad\qquad 1 \ 2 \ 1$$

$$111 \times 111 = \qquad\qquad 1 \ 2 \ 3 \ 2 \ 1$$

$$1111 \times 1111 = \qquad\qquad 1 \ 2 \ 3 \ 4 \ 3 \ 2 \ 1$$

$$11111 \times 11111 = \quad 1 \ 2 \ 3 \ 4 \ 5 \ 4 \ 3 \ 2 \ 1$$

This palindromic pattern continues all the way up to the number 111111111, which when squared comes to 12,345,678,987,654,321. To check this you may have to resort to calculating by hand, because your calculator is unusual if it displays more than 12 digits.

This symmetrical pattern might remind you of Pascal's Triangle, which cropped up in the previous chapter. In fact, there is an even closer connection between the number 11 and Pascal's Triangle. The first five lines of Pascal's Triangle are the powers of 11!:

$$11^0 = \qquad\qquad 1$$

$$11^1 = \qquad\qquad 1 \qquad 1$$

$$11^2 = \qquad\qquad 1 \qquad 2 \qquad 1$$

$$11^3 = \qquad 1 \qquad 3 \qquad 3 \qquad 1$$

$$11^4 = \quad 1 \qquad 4 \qquad 6 \qquad 4 \qquad 1$$

After that the pattern goes wrong because digits carry over, but deep down Pascal's Triangle and the powers of 11 are really the same thing. (By the way, if you are wondering about 11^0, it is a mathematical convention that any number raised to the power zero is 1, not 0.)

The beauty of ABC

Think of a three-digit number, any number you like, 832 for example. Now tap your number twice into a calculator (in my case I'd tap in 832832) to create a six-figure number. Since your number was plucked randomly from thin air, what would you say is the chance that it is divisible by the palindromic number 11? Fairly unlikely, you might think. But you will find that whatever number you started with, your six-figure number will always be divisible by 11. Furthermore, it is also divisible by 7 and also by 13, and if you divide the number by 11, 7 and 13 (in any order) you will find yourself left with a three-digit number – the same one that you chose at the start.

The reason why this works, by the way, is that 7 x 11 x 13 = 1001, and if you have a number of the form ABC and multiply it by 1001, you get ABCABC. Reversing the argument, that means that ABCABC divided by 1001 must be ABC.

Does every number make a palindrome?

As it happens, it is possible to turn just about any number into a palindrome. Take the number 142 for example. If you add it to its reverse, 241, you get 142 + 241 = 383 – a palindrome.

It doesn't always work so quickly, but if you repeat the process, a palindrome typically appears within a few moves. For example:

$$382 + 283 = 665$$
$$665 + 566 = 1{,}231$$
$$1231 + 1321 = 2552 \text{ – there's the palindrome}$$

The number 59 generates an even prettier result:

$$59 + 95 = 154$$
$$154 + 451 = 605$$
$$605 + 506 = 1111$$

The numbers 382 and 59 took just three steps to form a palindrome, and it is rare for numbers to take more than six or seven steps. But there are some odd exceptions. For example, the number 89 takes 25 steps before becoming a palindrome (see the table below).

step 1: 89 + 98 = 187
↓

step 2:   187
↓   + 781

step 3:   968
↓   + 869

step 4:   1837
↓   + 7381

step 5:   9218
↓   + 8129

step 6:   17347
↓   + 74371

step 7:   91718
↓   + 81719

step 8:   173437
↓   + 734371

step 9:   907808
↓   + 808709

step 10:   1716517
↓   7156171

step 11:   8872688
↓   + 8862788

step 12:   17735476
↓   + 67453771

step 13:   85189247
↓   + 74298158

step 14:   159487405
↓   + 504784951

step 15:   664272356
↓   + 653272466

nearly there...

step 16:   1317544822
↓   + 2284457131

step 17:   3602001953
↓   + 3591002063

step 18:   7193004016
↓   + 6104003917

step 19:   13297007933
↓   + 33970079231

step 20:   47267087164
↓   + 46178076274

step 21:   93445163438
↓   + 83436154439

step 22:   176881317877
↓   + 778713188671

step 23:   955594506548
↓   + 845605495559

step 24:   1801200002107
↓   + 7012000021081

step 25: = **8813200023188**

– a palindrome. Yippee!

And if you like big numbers, then how about 1,186,060,307,891,929,990 (that's about 1.1 quintillion in short-hand). It was discovered in 2005 that it takes a staggering 261 iterations for this huge number to reach its palindrome, a 119 digit number:

4456266587897643762243784897665387038888 47 8366259842585596343695585248952663874888 83 0783566798487342267346798785662654 4

Depending on your point of view, you might find it joyful that somebody actually made the effort to work this out – or just odd. Anyway, at the time of writing, nobody had found a number that forms a palindrome in more than 261 iterations, making this quintillion number the most 'delayed' palindromic number known to man, a fact surely worth recording in *The Guinness Book of Records* (for some reason it hasn't appeared there yet.)

But there could, conceivably, be a much smaller number that will beat this record. 196 might seem an innocuous little number, but astonishingly, nobody has yet found a palindrome despite millions of iterations. It is the smallest of many numbers that are now thought to be 'unpalindromable' (a word I just made up). These are known as Lychrel numbers. The so-called Palindrome Conjecture says that Lychrel numbers exist, but this is one of several mathematical problems that is easy to explain, yet has so far turned out to be impossible to prove.

Three mysterious numbers

There are plenty more 'symmetrical' number curiosities that you can explore with a calculator. For example, take the number 33^2, which is 1089.

If you multiply 1089 by 9, you reverse the number, to get 9801. Other multiples of 1089 also pair up into a pattern – the numbers in the middle form palindromes:

$$9 \times 1089 = \mathbf{9801} \quad \mathbf{1089} = 1089 \times 1$$

$$8 \times 1089 = \mathbf{8712} \quad \mathbf{2178} = 1089 \times 2$$

$$7 \times 1089 = \mathbf{7623} \quad \mathbf{3267} = 1089 \times 3$$

$$6 \times 1089 = \mathbf{6354} \quad \mathbf{4536} = 1089 \times 4$$

That just leaves 5 x 1089, which happens to form a palindrome all on its own: 5445

Incidentally, 1089 is also the basis of a popular 'mystery' that connects it to all three-digit numbers. Take any such number, 478 say, and reverse it, then subtract the smaller number from the larger. 874 – 478 = 396.

Now add this number to its reverse: 396 + 693 and the answer is 1089. The result is always 1089 for *any* three-digit number you choose (try it!), so long as the first digit differs from the last by at least two. So it will work for 265, but it won't work for 393. (Can you see why?).

Another pretty number is 12345679. Look at it carefully – it contains all of the digits except for 8. So what happens if you multiply the number by its missing 8? The result is:

12345679 x 8 = 98765432

If instead you multiply 12345679 by 9, you get our old friend

111,111,111 (though you are scuppered if your calculator only runs to eight digits). And if you want to surprise somebody, put 12345679 on your calculator, ask them what their favourite number is between 1 and 10, and then multiply 12345679 by *their number times 9*. So if they say '4', multiply by 36. The answer will be 444,444,444.

Finally, one of the most mysterious numbers of all is 142857. Multiply it by 7 and you get 999999, but if you multiply it by the numbers 1 to 6 you get:

$$1 \times 142857 = \mathbf{142857} \quad 142 + 857 = \mathbf{999}$$

$$3 \times 142857 = \mathbf{428571} \quad 428 + 571 = \mathbf{999}$$

$$4 \times 142857 = \mathbf{571428} \quad 571 + 428 = \mathbf{999}$$

$$5 \times 142857 = \mathbf{714285} \quad 714 + 285 = \mathbf{999}$$

$$6 \times 142857 = \mathbf{857142} \quad 857 + 142 = \mathbf{999}$$

Notice how the six digits in each of the products are always the same, but rearranged in a cycle. If the digits 1 to 8 are arranged in a circle, the numbers form a symmetrical pattern.

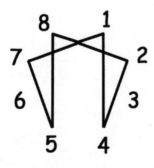

The mysteries of the number 142857 run deeper. Multiply it by any three-digit number you like, 649 for example:

142857 x 649 = 92 714 193

Split the answer into numbers with three digits, starting from the right, and then add these new numbers together.

92 + 714 + 193 = 999

In this case the answer is 999, and indeed most of the time the answer will be 999. If you're unlucky, it will be 999 x 2 (= 1998), and if you choose a number with more digits, then the result will always be 999 or a multiple of 999.

It is not a coincidence that certain numbers throw up appealing patterns like this. The way that mathematics works dictates that it will be so. But palindromes and other symmetrical numbers appeal to our love of pretty patterns, whether we're mathematicians or not.

8

Sudoku, and other magic squares

You will need: a pencil and an eraser.

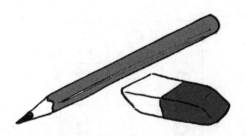

Back around the time when the ancient Greeks and Romans were getting excited about palindromes, a different pattern was attracting some attention thousands of miles to the east. Sometime in the first century AD, a Chinese philosopher called Tai Te was the first to describe a square pattern filled with symbols that had probably been discovered hundreds if not thousands of years before. It was a square called the Lo Shu:

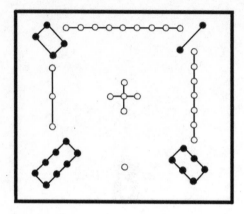

This curious pattern, so the story went, had been spotted on the back of a turtle as a sign of how many sacrifices to make after a particularly bad flood. Study it more closely and you will notice nine shapes each containing a different number of dots. If you count up the number of dots in each row or each column, the answer is always 15 – so whichever way the discoverers of the turtle counted, they would always end up with that number.

4	9	2
3	5	7
8	1	6

A square in which the numbers in each row and column add up to the same total is known as a magic square, and after the Chinese discovered these mysterious squares, an interest in them slowly spread across the world via India and Arabia.

One of the simplest ways to create a magic square is to take the numbers 1 to 3 and arrange them in a 3x3 grid so that each number appears only once in each row and column, like this:

1	2	3
2	3	1
3	1	2

Each row and each column adds up to six, though in this example only one of the diagonals adds up to six – the other adds up to nine. So this square is only a little bit magic.

This type of square, in which each symbol appears in each row and column exactly once, is generally known as a 'Latin square' thanks to a mathematician called Euler who first investigated them in detail. The reason why they became known as Latin squares was that he used Latin letters (rather than Greek letters) in his examples[3].

Sudokus

If the idea strikes you as familiar, it is because Latin squares have shot to international prominence since the invention of the Sudoku puzzle. A completed Sudoku is a Latin square, which not only has the property that every row and column has numbers that always add to 45, but the nine 3x3 squares within it also add to 45, too. Here's an example of such a square, which is the solution to a Sudoku puzzle:

---

[3] Latin squares are nothing to do with the Romans, though the AREPO ROTAS TENET SATOR OPERAS square on page 93 is a square made up of Latin, and is almost a Latin square.

8	5	7	4	9	3	1	2	6
4	9	1	7	2	6	8	3	5
2	6	3	8	5	1	9	7	4
7	8	6	2	3	5	4	9	1
3	1	4	9	6	8	7	5	2
9	2	5	1	4	7	3	6	8
1	3	8	6	7	2	5	4	9
6	7	9	5	8	4	2	1	3
5	4	2	3	1	9	6	8	7

To invent a Sudoku puzzle of your own, all you have to do is remove a few of the numbers in the grid above. The solver then has to work out what numbers you removed. (Actually, it's not quite as simple as that – if you want to create a good Sudoku, that is.)

There are many who like to claim that Sudoku puzzles have nothing to do with mathematics. For years, *The Independent* newspaper in the UK has put a note next to each day's puzzle to say that 'There is no maths involved'. What they probably mean is that no *arithmetic* is involved, since it isn't necessary to add or subtract any numbers to solve the puzzle. The truth is that Sudokus are deeply mathematical, since solving the puzzle require rigorous logic and reasoning.

There is plenty of maths in setting a good Sudoku puzzle, too. For example, which digits should you remove from a completed grid in order to make a good Sudoku? It is by no means obvious. If you take away too few, the puzzle is trivial, but if you take away too many, the puzzle may become too hard to solve by pure logic, and may even turn out to have more than one solution. To take this to an extreme, if you were to remove

80 of the 81 numbers in the filled Sudoku above, leaving just one digit, there are hundreds of millions of valid solutions to the puzzle.

You might be curious to know how many digits there have to be in the grid before a Sudoku can have a unique solution. The answer, it turns out, is 17. Nobody knows why it is 17, and it hasn't yet been conclusively proved that 17 is the minimum, but there are several examples of Sudokus with unique solutions that have only 17 entries, and none with 16. Here is just one of them.

9		8						
				7				
			5	1				
	1							
					2	9		
			6		9	2	8	
	5	1						
					4		2	
							3	

If you feel tempted to have a go at solving it, be warned: it is extremely hard, though it can be solved using logic without the need to resort to trial and error. That will please the Sudoku purists who don't like puzzles where you can't just use logic alone.

Most Sudoku puzzles have rather more than 17 digits in them at the start, with 25 being a typical number. It has been worked out that there are over five billion distinct Sudoku patterns that have unique solutions, so newspapers aren't going to run out of new puzzles any time soon.

However, Sudoku experts will know that some Sudoku puzzles are better than others, and Sudoku puzzle-setting can be as much an art as a science. For example, look at this example, which is taken from a national newspaper:

7				6				2
			9		1			
8		5				6		7
	4			5			6	
5			3		2			8
	2			8			7	
9		3				5		1
			8		3			
6				4				9

Do you notice anything curious about it? It is easier to spot if the numbers are replaced by blobs:

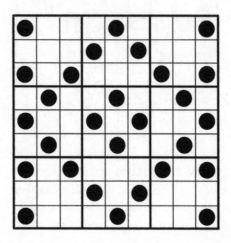

The blobs form a beautiful symmetrical pattern – unchanged if you turn the page upside down or view it in a mirror. Why does it have this pattern? Because the setter thought it looked more aesthetically pleasing. The symmetry of a Sudoku has nothing to do with how easy or hard it is to solve, so this is a case of the mathematician wanting to express himself.

Newspapers and magazines across the world are split between those that always carry symmetrical Sudokus (*The Daily Telegraph* and *Times* for example) and those whose Sudokus are asymmetrical (*The New York Times*, *The Independent*). The smallest number of digits in a symmetrical Sudoku pattern is believed to be 18, if the solution is to be unique.

Sudokus are designed for puzzling rather than for the 'ooh, aah' of a magic square. But there are variations of the Latin square that can be very magical indeed.

Four-by-four magic squares

Simply increasing a grid from 3x3 to 4x4 can introduce a whole new layer of intrigue to a magic square.

Take this grid of numbers, for example:

10	7	8	11
4	1	2	5
12	9	10	13
7	4	5	8

I'm thinking of a secret number, and I'm predicting that you are going to select four numbers from the grid at random, and your four numbers will add to my predicted number. (If you don't want to deface this book, I suggest you make a copy of the number grid.)

Choose any number in the grid, and put a circle around it. You have absolutely free choice, remember. Once you have selected the number, cross out all of the other numbers in its row and all the other numbers in its column. (So if you chose the '9', cross out the 7, 1 and 4 in its column, and the 12, 10 and 13 in its row.)

Now choose another number from the grid that hasn't been crossed out, and put a circle around it. Again, cross out the other numbers in its row and column. Next choose a third number and circle it, crossing out the others in its row and column.

You should find that there is now only one number remaining that hasn't been circled or crossed. Put a circle around that number.

You now have four numbers with circles around them. There is no possible way that I could know which four numbers you were going to choose. And yet, I confidently predict that the four numbers you have circled add up to... **29**.

Am I right? How does it work?

There is something similar to a Latin square principle at work here. To create the 'magic' grid, I started by choosing eight numbers that add to 29, and placing along the top and the side of the grid, like this:

The grid can be filled in by adding the number at the top of the column to the number at the side. For example, the top left corner is 4 + 6 = 10. When you circled your numbers, you were forced to choose exactly one number from each row and one from each column, so whichever selection you made, it must have been a combination of

4 + 1 + 2 + 5 + 6 + 0 + 8 + 3 which comes to 29.

Most of the secrets behind the creation of magic squares have been known for centuries, and there are some easily learned techniques for building a magic square of your own. One of the simplest techniques produces a remarkable four by four magic square. To create the square make a blank grid and lightly mark the diagonals. Then fill in the numbers 1 to 16 along the rows of the grid starting in the top left corner:

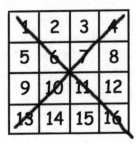

Now rub out all the numbers that don't appear on a diagonal.

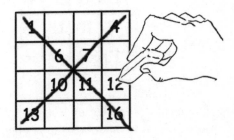

Next fill in the square again, but this time from 16 to 1, only writing down the numbers that don't lie on diagonals:

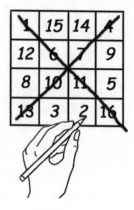

The result is a magic square with some remarkable properties. Every row and column adds to 34. So do the diagonals. And the four corner squares, as well as the four squares in the centre. The four squares forming the top left quarter of the grid add to 34, as do those in the other three corners of the grid. So do the middle two squares on the left plus the two on the right. And several other combinations besides. In fact there are over 30 symmetrical ways of making the number 34.

1	15	14	4
12	6	7	9
8	10	11	5
13	3	2	16

The middle numbers on the top line form 1514. That is the year when Albrecht Durer made a famous engraving, in which this magic square (or one very like it) appears cryptically in the corner. It shows that magic squares are one of those things that can intrigue artists and mystics just as much as mathematicians.

The ultimate symmetrical 4x4 square

The 1514 square is a nice historical example of a magic square, but my favourite magic square has a much more modern look. It is the 176 Square (the 176 happens to be the number of my local bus, but that isn't why I like the square). I have written the numbers in 'calculator' form for a reason:

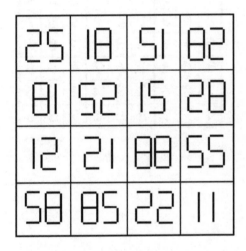

The rows, columns and diagonals add to 176, as do numerous other symmetrical groups of four around the square, just as with the earlier 4x4 square. But there is an added twist. If you now turn the book upside down, you'll discover that the magic square has now changed, but it still works, and still adds to 176. And if you look at it in a mirror, it produces yet another magic square that again adds to 176.

Cool, huh?

Solution to the 17-Sudoku puzzle.

9	7	8	4	2	6	5	1	3
1	3	5	9	7	8	6	4	2
6	2	4	5	1	3	8	7	9
2	1	9	7	8	5	3	6	4
8	6	3	1	4	2	9	5	7
5	4	7	6	3	9	2	8	1
3	5	1	2	6	7	4	9	8
7	8	6	3	9	4	1	2	5
4	9	2	8	5	1	7	3	6

Let's take the short cut

You might want: A large bar of chocolate

A farmer is standing at a gate one day, admiring his flock of sheep, when a tourist comes along to join him.

'Those your sheep?' asks the tourist.

'Aye, they are,' says the farmer. 'And how many do you reckon there are out there?'

The tourist pauses for a few moments, then replies: 'Three hundred and eighty-six.'

The farmer's jaw drops. 'How on earth did you know that?'

'Easy,' says the tourist, 'I counted the legs and divided by four.'

The point of this, of course, is that there's always more than one way of doing things, but some are more efficient than others. The same applies in maths. Anyone who taught you that there's only one way of doing multiplication, say, was grossly misleading you. There are thousands of different ways, some simple but slow, some complicated, some plain weird. But one of the things that send tingles down the spine of a mathematician is discovering a short cut. Mathematicians will often use the word 'elegant' when describing something mathematical, and by elegant they often mean a solution to a problem that is clear, logical and brief.

The train puzzle

One vintage example concerns two trains that are approaching each other on the same track, each travelling at 50mph. When the trains are 100 miles apart, a fly that always flies at 60mph, sets off from the front of Train A directly towards the oncoming Train B. When it gets to Train B it turns around and heads straight back towards A, and it continues to fly back and forth between the two trains until CRUNCH! the two trains smash into each other, crushing the fly in the process.

How far did the fly fly – before it met its unfortunate end? The natural way to tackle this question is to work out how far the fly travels on each leg of the journey. The first leg will be slightly less than 100 miles, the return leg will be less again, so if we can just work out a formula, we can add all those legs of the journey together and get the answer.

There is, however, an alternative. Let's ignore the journey of the fly for a moment. At the start of the story, the trains are 100 miles apart, and both are travelling at 50 miles per hour. That means that in one hour, the trains are going to crash (each will have travelled 50 miles).

And in that hour, what happens to the fly? We are told that the fly always travels at 60mph. This means that in the hour that it takes the trains to collide, the fly must travel exactly 60 miles. It doesn't matter if the fly flies in a straight line, zigzags or loop-the-loops, the answer will be the same.

The knockout and the chocolate bar

An equally ingenious solution exists for the so-called knockout tournament problem. Imagine there are, let's say, 64 teams taking part in a football knockout tournament. In every match in the tournament, one team knocks out the other, until there are only two teams left for the final, and the winner of that match gets the trophy. How many matches are there in the tournament?

In the first round, if there are 64 teams then there must be 32 matches. In the next round there are 32 teams, and they play 16 matches. Following this pattern, it becomes clear that the total number of games is 32 + 16 + 8 + 4 + 2 + 1 which when added together comes to the answer 63.

What if 97 teams start the tournament? Now some teams are going to need to play more games than others (usually this happens by having some teams get 'byes' in the first round). You might like to try working out how many matches are going to be needed, first by working out how many teams get byes.

Alternatively, you might just want to know the short cut. The secret to this problem is in the way you ask the question.

People usually start from the point of winning: how many matches get won in each round?. But since in every game there is a winner and a loser, it is just as fair to concentrate on the losers, who get knocked out. How many of them are there? Well, by the end of the tournament, everybody has been knocked out except for the final winner. Each team gets knocked out by one match. Therefore, the number of matches will always be exactly one less than the number of teams, no matter how you structure the tournament. If there are 64 teams, there will be 63 matches.

All of which appears to have nothing to do with the chocolate bar challenge. Suppose you have a monster slab of chocolate, with 40 chunks, like this:

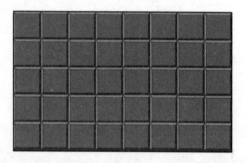

It just so happens that you have 39 friends with whom you'd like to share the chocolate, so you decide to 'snap' the chocolate into pieces along the horizontal and vertical lines on the bar. Each snap involves taking one rectangle of chocolate and snapping it into two pieces along one of the lines. You are not allowed to 'piggy back' two or more pieces to break together, as this may mean you get an uneven break.

The question is, what is the most efficient way to snap the large slab into 40 individual square chunks? And how many snaps does it take to do it?

(At this point you may wish to find yourself a real bar of chocolate and experiment – rewarding yourself every time you create a single chunk.)

Some people begin to devise strategies when solving this problem, such as *'always attempt to divide a piece of chocolate into equal parts, if possible'*.

However, it turns out that no such 'strategy' is necessary. That is because, whatever method you use, it will always require exactly 39 snaps to create 40 individual chunks. Why? Because each snap always generates one extra piece of chocolate. At the start you have one (very large) piece of chocolate, and at the end you have 40 pieces. That means it must have taken 39 snaps to create them, whatever approach you used. The number of snaps needed is always one less than the number of chunks you want to end up with.

Does this sound familiar? This is *exactly* the same problem as you encountered just moments ago with the knockout football tournament. Chocolate snapping and football knockouts may sound very different, but the underlying maths is just the same. That's another beauty of maths – that the same solution can often be used in apparently very different situations.

King Tutu and the babies

In earlier chapters we've already seen how probability can throw up some complex calculations. The story of the island of Tutu, a tale of grim dictatorship and a quest for male supremacy, is one such example.

It all happened many years ago. King Tutu lived on the island of Tutu, an island that, like every other place on earth, consisted of 50 per cent males, and 50 per cent females. But King Tutu was worried. He had heard that the neighbouring island of Ping Pong was growing restless, and rumour had it that one day the Ping Pongians would build a fleet so that they could launch an invasion. Anxious to defend his island, King Tutu decided he would need more warriors, and that meant that he wanted more men on his island. So he introduced a new Law, decreeing that women were allowed to have as many boy children as they

wished, but *as soon as they had a girl child, they would be permitted no more children.* Some mothers who wanted lots of children were lucky. There was one who had 10 boys! But there were others whose first child was a girl, and they were condemned to raise just one child, whether they wanted more or not.

Thirty years later, there had been no invasion from Ping Pong. But there were lots of young people on the island. King Tutu decided to hold a census of everyone under the age of 30. Weeks later, the Census clerk hurried into the King's palace bearing the news. 'Sire, we have counted all 1,200 of your young people. There are 245 farmers, 97 bakers...' 'Enough!' cried the King. 'There is only piece of information that I want. Of the 1,200 young people, how many are male and how many are female?'

Can you predict the answer? What is your intuition telling you – more males or more females?

Both answers are possible – it's possible that every baby born in the 30 years was a boy, and it's conceivable (excuse the pun) that all were girls, born into single child families. The actual result will be somewhere between the two, but where should we expect it to be?

The surprising – and counter-intuitive – answer is that the split of males and females on the island remains around 50–50. In other words, the King's draconian policy had **no effect whatsoever** on the split between boys and girls.

It's possible to work out this answer by calculating the probabilities of the different possible combinations of children, but that involves some treacherous maths. Fortunately there is an elegant short cut that requires no calculations whatsoever, if you just imagine yourself working in the maternity unit of Tutu hospital.

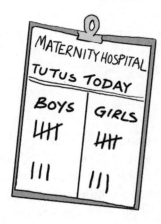

Imagine a typical morning in Tutu hospital. A pregnant woman arrives, goes through labour, and out she comes with her beautiful baby. Of course we all know that a baby has an equal chance of being a boy or girl, and this new child can be either. If it is a girl, then this mother will never be allowed to have more children. But that is irrelevant.

The next woman comes in, then the next and the next. Each time, the new baby is either a boy or a girl, with equal chance, and whether it's a first time mum or a mother of 10 makes no difference, babies will always arrive in the ratio one boy to one girl. So the proportion of boys and girls in the population *will not change!*

Since babies are randomly boys or girls with a 50–50 split, the maths of babies is very similar to the maths of tossing coins discussed in Chapter 6. This means that exactly the same curious patterns that arise with Heads and Tails arise in families.

For example, if you have two children, the chance that both are boys is the same as tossing two Heads, in other words 25 per cent. The chance of having exactly one boy is the same as tossing a Head and a Tail (in either order) which is 50 per cent. And as with unbiased coins, if you have four boys in a row, the chance that your fifth child will be a boy – like the chance of a fifth Head – is still 50 per cent. Those who claim that their family is renowned for producing all boys might be quoting what has already happened, but it has no influence on what will happen in the next generation.

The Ant puzzle

What connects all of the problems so far in this chapter – the fly, the knockout tournament, the chocolate and Tutu island – is that a problem that at first glance seemed complicated turned out to have a simple solution. The ability to see the wood for the trees is one of the most valuable skills you can possess, and a quality that has marked out many great leaders in history.

Let's conclude with what is perhaps, at first glance, the most complicated problem of the whole chapter. It concerns some ants.

These ants behave in a particular way. They always walk in straight lines, and always at a constant speed of 1 metre per minute (which makes them rather slow, for ants). If they ever bump into another ant, they immediately turn around and walk in the opposite direction. Finally, if they ever walk to the edge of a surface, they fall off like lemmings.

Now for the problem. There are 100 of these ants, and they are carefully sprinkled along the edge of a 1 metre-long ruler which is raised above the ground. Some of the ants are pointing to the left, some are pointing to the right, but it is all quite random. The ruler is so narrow that there are no passing places, so every time two of the ants meet they turn around. And when they reach the end of the ruler, they drop off the end.

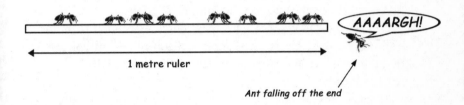

AAAARGH!

1 metre ruler

Ant falling off the end

The question is: how long must you wait before you can guarantee that ALL of the ants have fallen off the end of the ruler?

Before attempting to work it out, just think about how complex the situation is. At the beginning collisions between ants will be happening almost constantly – unless by complete fluke all the ants were pointing in the same direction from the start. It is hard enough trying to follow what would happen to the ants if there were three of them, let alone 100.

An astrophysicist was shown this problem, pondered it for a second and (being exceptionally bright, and used to complicated problems) he immediately started to think about building a sophisticated simulation that would enable him to work out the patterns in the millions of different ant scenarios.

Yet all of that is unnecessary, with one crucial 'Aha!' bit of insight. Think about what happens when two ants collide. In the instant before the collision, one ant is walking to the right at 1 metre per second, and one is walking to the left at the same speed. After the collision (when the ants reverse) there is one ant walking to the right and one to the left at the same speeds. In other words, *nothing has changed*, at least if you blur your eyes a little. What this means is that, mathematically, two ants colliding is identical to two ants walking past each other. We might as well regard this problem as 100 ants each in their own private lane, walking left or right, and never bumping into anything.

In which case, the longest distance that an ant will have to walk is the full length of the ruler, which is 1 metre, and the longest time you have to wait for all the ants to fall off the ruler is therefore one minute. It is as 'simple' as that.

We started the chapter with flies flying and ended with ants walking. In both cases, the secret – the genius – in solving a problem was the ability to step back and look at the big picture, not the minute detail.

That is why 'How far did the fly fly?', and 'How long did the ants walk?' are two insect questions that can get mathematicians very excited.

Journey to the centre
of the triangle

You will need: A pencil, a ruler and an atlas.

I once had a geography teacher who liked to draw countries as triangles. It speeded up his blackboard sketching no end. His map of Great Britain looked like this:

And his map of the USA looked like this:

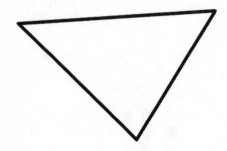

As far as I recall, that teacher never asked us to locate the centre of Britain, which is a shame since his triangle map might have helped to resolve a dispute that arose in 2005 between the London store Harrods and a village pub in the North of England.

The argument started when Harrods announced that they were holding a raffle, with the first prize being a small plot of land that was 'the centre of Britain'. It turned out, rather conveniently, that the plot of land was a patch of boggy turf in the middle of rural Lancashire, near the village of Dunsop Bridge, of little practical value to anybody.

However, this did not please the owner of a pub 71 miles further north in the town of Haltwhistle. His pub was called The Centre of Britain, and any claim that the centre might be somewhere completely different would clearly be bad for business.

So who was right?

We can begin to understand the thinking behind the two claims by trying to locate the centre of a triangle. Perhaps the most obvious way to find the middle of something is to simply measure 'halfway along' and 'halfway up'. This, after all, is how you would find the middle of a rectangle – and of course halfway along and halfway up is where you'd find the centre spot of a football pitch.

But halfway along and halfway up doesn't work quite as well with triangles. Take the Great Britain triangle for example:

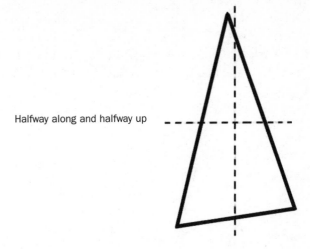

Halfway along and halfway up

Sure, the point where the lines cross does look sort of 'middling', but what happens if we tip the triangle over onto its longest side?

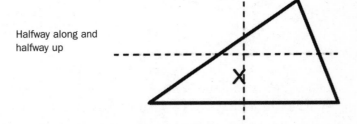

Halfway along and halfway up

The centre of 'Britain', previously at the point x, has moved!

The halfway up/halfway across technique is the one that was used to locate the centre of Britain in Haltwhistle, but if the location of the centre depends on which way you hold the map, it does suggest that its case for centrality is, to say the least, a little tenuous.

The surveyors who chose Dunsop Bridge as the centre of Britain reckoned they were on stronger ground (even if the ground was boggy), because they were using the centre of gravity method. Take a triangle and draw a straight line from each corner of the triangle to the centre of the opposite side (marked with an X), like this:

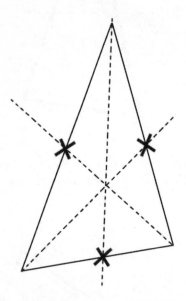

The three lines always coincide at a point, whatever shape the triangle – and whichever way you hold it up. This point is sometimes called the 'centroid', but is better known as the centre of gravity. If you cut out the triangle and rest the centroid on the point of a needle, the triangle should balance perfectly. The centroid is the only point on the surface of the triangle where this will work, and it makes sense that this point of balance would make a sensible point for the centre of a country.

There is, however, a snag. Not all countries resemble triangles. Suppose a country were L-shaped like this:

If you try to find the point where this shape will balance on a pin, you will be disappointed. That's because the centre of gravity in this case is outside the border of the country – roughly at the point X. A rule whereby a country's centre might be outside its borders makes no sense at all!

Fortunately there is yet another method that guarantees to find a centre that always remains fixed and is always inside the borders. This method is to find the place that is *the furthest from the border* (or in the case of an island, the furthest point from the sea).

For a triangle, this point is easy to find. From each corner, draw a line that splits the angle in two.

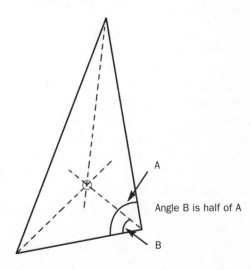

Angle B is half of A

If you extend the three lines, you will find that they always meet at a point. This is a rather neat – perhaps even surprising – piece of geometry. After all, there's no obvious reason why the lines should always meet. This meeting point is known as the triangle's incentre, and it is usually a different point from the centroid. (In the case of our 'Britain' triangle above, the incentre is further south than the centroid.) The incentre happens also to be the centre of the largest circle that you can draw inside the triangle – and if you think about it, that also makes it the furthest point from the 'sea'.

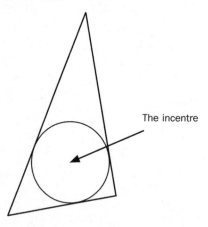

The incentre

The largest circle that can be drawn inside the map of Britain has its centre somewhere in the Midlands, many miles from Dunsop Bridge and Haltwhistle. Unfortunately its precise location is disputed just as much as the merits of the other two locations – because it depends on what you define as being the sea.

Even more problems arise when trying to locate the centre of the USA. The centre of gravity of the original 48 states is in a place called Lebanon in Kansas. But the USA is so big that its curvature with the earth shifts the location by about 50 miles. When Alaska and Hawaii are thrown in, the centroid moves up to North Dakota, and the halfway up, halfway across centre goes down to Texas. So nobody takes the centre argument seriously in the USA – except for the tourist board in Kansas.

The theory of triangles is one thing, but the realities and politics of geography are quite another.

Napoleon's Theorem

Politics and triangles have been unlikely bedfellows before. One example was the French emperor Napoleon Bonaparte, famous as a great military strategist, but less well known as being a lover of geometry. Napoleon was particularly interested in triangles. This shouldn't be too surprising, since in his day an understanding of geometry was critical for any military leader (to be able to work out the range of a cannon, for example).

There is a triangle theorem that Napoleon is said to have discovered. Take any triangle, as irregular as you like.

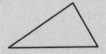

Then draw an equilateral triangle on the full length of each side.

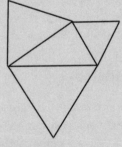

Now find the centres of each of the three equilateral triangles.

Because the three triangles are all equilaterals, it doesn't matter which 'centre' you choose – in equilaterals the incentre and centroid are in the same place.

Finally, join the three centres together to form a new triangle.

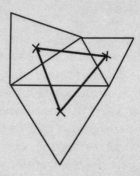

No matter how irregular the triangle that you started with, the new triangle will also be an equilateral. This is Napoleon's theorem.

If Napoleon really was the first to formally prove this, it was some nifty mathematics on his part. Whether he did or not, it bears his name to this day. You can see where those ingenious military tactics came from....

Right-angled triangles

The centre points of a triangle are not the only feature of this simple shape that might interest a map-maker. The most important triangle in the world of maps is the right-angled triangle, because it can be used to work out distances between any two points on a grid. The technique in question is the theorem of Pythagoras, which enables you to work out the distance from A to C so long as you know the distances AB and BC (which can be easily worked out from the co-ordinates).

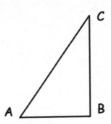

If you want an example of just how creative maths can be, then look no further than Pythagoras' theorem. In case you have forgotten what this is, here is a quick refresher. Take any right-angled triangle:

Now draw a square on each side of the triangle:

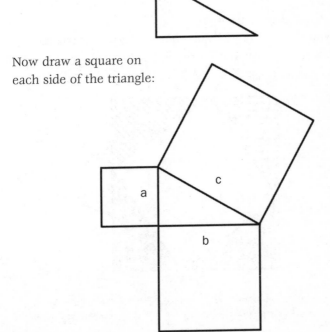

The theorem says that the area of the square on the longest side (known as its hypotenuse[4]) is exactly the same as the total area of the squares on the other two sides. In shorthand, $a^2 + b^2 = c^2$

So where's the creativity in that?

The creative part is that over the centuries, mathematicians have been incredibly imaginative in finding different ways of *proving* it. Indeed there are known to be at least 367 distinct proofs of Pythagoras' theorem. We know this because more than a century ago, a chap by the name of Elisha Scott Loomis wrote a book that described every one. One of those proofs came from none other than President James Garfield of the USA, who is remembered for little else because after a few months in office somebody decided to shoot him.

When there are hundreds of different proofs to choose from, the temptation is to ask: *which one is the best?* There is one proof that I think stands out because it is visual and intuitive, requiring no numbers at all.

You can call it the carpet proof.

Imagine you have a square room, and four identical triangles of carpet that fit perfectly around the edge of the room, leaving a square of bare floorboards in the middle like this:

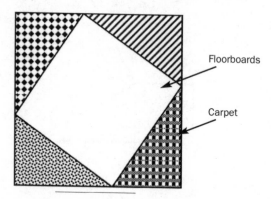

Floorboards

Carpet

[4] Hypotenuse means 'under tension', and presumably refers to the fact that the longest side of a triangle looks a little like a piece of string stretched between the ends of the other two sides.

The triangular pieces of carpet are all right-angle triangles, and the square of floorboards in the centre is the 'square on the hypotenuse' of each triangle.

Now, let's shift the carpet around. We'll take the piece in the bottom left and slide it diagonally up to join the piece that's top right. And we'll take top left and bottom right pieces and slide them so that they join up in the bottom left corner. Like this...

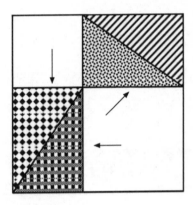

Now there are *two* uncovered patches on the floor. Both of them are squares. But wait a moment. The sides of the two squares are the same as the two shorter sides of the triangular pieces of carpet. And since the amount of uncovered space on the floor has not changed, we've shown that the area of the large square must be the same as the area of the other two squares of the other two sides. That's all Pythagoras was trying to say.

Cutting triangles into pieces

Not only can all geometrical shapes be reduced to triangles, but triangles themselves can always be broken down into smaller versions of themselves. In fact it is always possible to divide a triangle into exactly four identical miniature copies of itself. The easiest triangle for seeing this is the right-angled triangle:

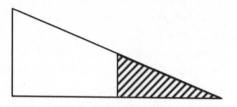

Draw a vertical line halfway along the base of the triangle. The shaded triangle that you create is a replica of the original, but one quarter the size. The remaining body of the triangle can now be divided into three identical triangles.

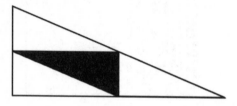

Notice that the black triangle in the centre is the same as the others, but has turned upside down.

You can experiment with other triangles to demonstrate to yourself that any triangle, however irregular, can be divided into four miniature versions of itself in this way, with the 'middle' black triangle always a flipped version of the others:

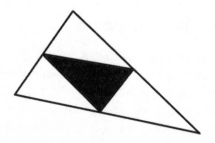

You can take this further by subdividing the white triangles into yet smaller replicas of themselves. If you start with a large equilateral triangle and keep dividing the white triangles into four, colouring the central triangles black, you end up with this:

This pretty pattern is known as the Sierpinski gasket, and it is a fractal. In other words, if you were to zoom into any part of the image you would find the same pattern repeating itself ad infinitum. (The Dragon Curve on page 60 was another example of a fractal).

Pascal's Triangle and Sierpinski's gasket

Sierpinski's gasket has a connection to a triangle that we encountered in chapters 6 and 7, Pascal's Triangle.

Below are the first few rows of Pascal's Triangle. Notice that some of the numbers in it are odd, and others are even.

```
                        1
                      1   1
                    1   2   1
                  1   3   3   1
                1   4   6   4   1
              1   5  10  10   5   1
            1   6  15  20  15   6   1
          1   7  21  35  35  21   7   1
        1   8  28  56  70  56  28   8   1
      1   9  36  84 126 126  84  36   9   1
    1  10  45 120 210 252 210 120  45  10   1
  1  11  55 165 330 462 462 330 165  55  11  1
1  12  66 220 495 792 924 792 495 220  66  12  1
1 13 78 286 715 1287 1716 1716 1287 715 286 78 13 1
1 14 91 364 1001 2002 3003 3432 3003 2002 1001 364 91 14 1
```

If you now blot out the even numbers with a thick black pen, something curious begins to emerge:

Blur your eyes, and the pattern of black blobs form triangles. This is none other than Sierpinski's gasket once again. It seems quite remarkable that a triangle that is made up of numbers should have this deep connection with another triangle that was purely geometrical.

If you were to keep going with Pascal's Triangle, to the millionth or the zillionth row of numbers, what proportion of the numbers in the triangle will be odd?

The answer to this can be found in the final chapter on infinity (If you want a sneak preview, you will find it on page 164.)

The Golden Triangle

We've just seen how any triangle can always be divided up into exactly four miniature copies of itself.

There is, however, one triangle that can also be divided into five copies of itself. The triangle in question is a very simple one, a right-angled triangle with one of the short sides double the length of the other:

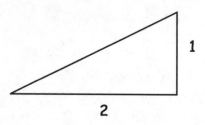

This can be divided into four triangles in the usual way:

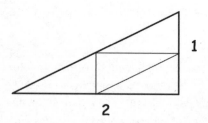

But this time, there is also way of dividing up the triangle into five identical miniature versions of itself, like this:

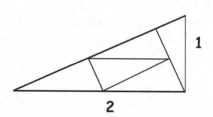

The shorter sides of the small triangles are still in the same ratio of 2:1. Indeed, to use language that my five-year-old daughter would relate to, it's as if this special baby triangle can be used to build a daddy triangle and a mummy triangle. All three triangles are the same shape, and Daddy – Mummy = Baby[5].

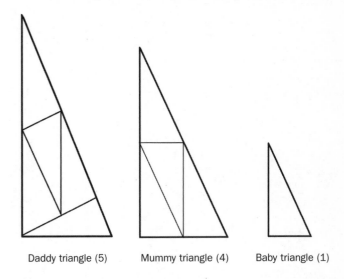

Daddy triangle (5)   Mummy triangle (4)   Baby triangle (1)

What is it about this particular 'baby triangle' that permits it to have a mummy and a daddy, rather than just a mummy?

The answer has something to do with the length of its longest side. As President Garfield could have told you, if we call the length of the hypotenuse 'Fred', then

Fred^2 = **2**^{2} + **1**^{2}

So **Fred**^{2} = **5**

And hence **Fred** = $\sqrt{5}$, approximately **2.236**

---

[5] As opposed to biological maths where Daddy + Mummy = Baby.

Remember that number. It's closely related to the meaning of life. Or so some people think, anyway. We'll meet it again in the next chapter.

Limericks, Leonardo and the number five

You might want: a book of limericks, a tape measure, a compass, a calculator and an apple.

'There once was a lady called Bright
Who travelled much faster than light
She woke up one day
The usual way
And came back the previous night'

Everyone loves a limerick. But perhaps there's a reason for that. Embedded within the classic limerick rhythm is a number pattern that has profound links with nature. The pattern has even been linked with great art and sculpture from the time of the ancient Greeks. It's a pattern that some regard as so beautiful it is almost divine. And it is all connected to the number five, which happens to be the number of lines in a limerick.

Here's the limerick again, reduced to a series of Dees and Dums. It goes:

> Dee Dum dee dee Dum dee dee Dum
> Dee Dum dee dee Dum dee dee Dum
> Dee Dum dee dee Dum
> Dee Dum dee dee Dum
> Dee Dum dee dee Dum dee dee Dum

Notice anything? Probably not. Don't worry, all will be revealed later.

Fibonacci

The mystery begins with a curious puzzle that was first set out in the year 1202AD by a man called Leonardo of Pisa, who was nicknamed Fibonacci. (Fibonacci means 'the son of Bonacci' – Bonacci being his dad).

Fibonacci wrote a book called *Liber Abaci* (A Book of Calculations) which introduced Europeans to the idea of the modern number system that was already popular in India and across the Arab world. It is partly thanks to Fibonacci that today we calculate using the numbers 0 to 9, rather than still being lumbered with Roman numerals.

Fibonacci's book was full of 'word problems', which were really just an excuse to show how to use the new-fangled number system. The puzzles were not easy reading. They included this one about rabbits (I paraphrase somewhat):

'A man has a pair of adult rabbits. Adult rabbits mate every month, and one month later they always give birth to one boy and one girl rabbit. Baby rabbits take one month to become adults, and then mate immediately. How many pairs of rabbits are there after one year?'

Using Fibonacci's unlikely set of rabbit-breeding rules, the number of pairs of rabbits grows as follows:

Start	**1 pair of adults**	**= 1 pair**
1 month	1 pair of adults, 1 pair of babies	= 2 pairs
2 months	2 pairs of adults, 1 pair of babies	= 3 pairs
3 months	3 pairs of adults, 2 pairs of babies	= 5 pairs
4 months	5 pairs of adults, 3 pairs of babies	= 8 pairs
5 months	8 pairs of adults, 5 pairs of babies	= 13 pairs
6 months	13 pairs of adults, 8 pairs of babies	= 21 pairs

7 months   **21 pairs of adults, 13 pairs of babies**   **= 34 pairs**

8 months   **34 pairs of adults, 21 pairs of babies**   **= 55 pairs**

9 months   **55 pairs of adults, 34 pairs of babies**   **= 89 pairs**

10 months   **89 pairs of adults, 55 pairs of babies**   **= 144 pairs**

11 months   **144 pairs of adults, 89 pairs of babies**   **= 233 pairs**

12 months   **233 pairs of adults, 144 pairs of babies** **= 377 pairs**

So the answer to Fibonacci's question was that at the end of the year there would be 377 pairs of rabbits. Fibonacci pointed out that in order to work out the numbers of rabbits each month, you simply add together the totals from the previous two months together. (1 + 2 = 3, 2 + 3 = 5 and so on). But so what? It's a good question.

Nobody knows why Fibonacci created this implausible puzzle. In case you aren't up with rabbit breeding, the truth is that rabbits cannot breed until they are several months old, they generally have litters of six or more, and unlike Fibonacci's rabbits, wild rabbits do not breed with their siblings as that would be genetic suicide.

Perhaps Fibonacci's story was simply a contrived way of creating a particular sequence of numbers that he already knew about.

This innocent looking sequence goes **1 2 3 5 8 13**...

Fibonacci made no comment about the significance of this sequence (perhaps he didn't realise there was any) and it went largely unnoticed for centuries, but gradually people started to discover it for themselves, and to notice that it has some unusual properties.

For example, if you pick any number in the sequence and square it, and then take its two neighbours and multiply them together, the answers always differ by one.

$$3 \times 3 = 9 \qquad 2 \times 5 = 10$$

$$8 \times 8 = 64 \qquad 5 \times 13 = 65$$

Numbers differ by 1

Or take our old friend Pascal's Triangle. Fibonacci numbers appear there too. (To be honest, Pascal's Triangle is such a treasure trove you can find just about any pattern that you could possibly imagine.). To find Fibonacci in Pascal, turn it on its side, and then draw vertical lines through the numbers:

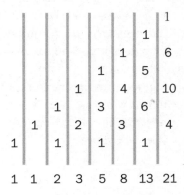

Add up the numbers in each column and they are Fibonacci.

The rabbit example that Fibonacci used to reveal the number sequence might have been unnatural, but Fibonacci numbers do genuinely crop up in the natural world. Many flowers that you will find in your garden, including those on trees, have a Fibonacci number of petals, with five the most common number. Slice an apple through its centre and you will find that the seeds make a five pointed star – not the first time that the number five has cropped up in this chapter, and it isn't the last either.

The number of spirals around pineapples and pine cones (barring freaks) will *always* be Fibonacci numbers, too (typically 8, 13 and 21).

Why do plants choose Fibonacci numbers in this way? Because it turns out that they are the most efficient way of packing the 'scales' around the surface.

There is also a curious connection between Fibonacci numbers and male bees, thanks to the particular way in which bees reproduce. Not all of the eggs that are laid by a queen bee are fertilised by males. The eggs that are fertilised become females (most of which become *workers*). The unfertilised eggs become males (known as *drones*).

Thanks to this method of reproduction, every male bee has only one parent (mum), whereas every female has two (mum and dad). This means that if you take any male bee, you can track back its ancestors as follows:

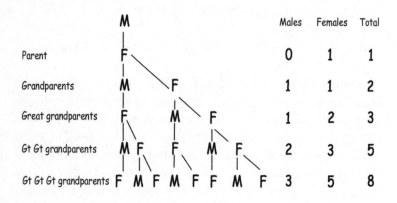

	Males	Females	Total
Parent	0	1	1
Grandparents	1	1	2
Great grandparents	1	2	3
Gt Gt grandparents	2	3	5
Gt Gt Gt grandparents	3	5	8

Notice anything interesting as you go back through the generations? Each number of ancestors is a Fibonacci number, and
because the mechanism for working out the number in the next
generation is (as it happens) the same as for Fibonacci's rabbits,
the pattern continues indefinitely. Well, not quite. If it really did
go on for ever, then a bee's number of ancestors would keep growing forever too, whereas the truth is that as you go back in time
the number of ancestors should get smaller, not bigger. (Like
humans, modern bees developed from just a couple of ancestors).

The explanation for this apparent paradox is that as you go
back through the generations, most of the ancestors will turn out
to be brothers and sisters with the same mother, so almost all of
the ancestors are being double-counted at very least.

Still, it's interesting that, in theory, bees could be Fibonacci
creatures.

The Lucas sequence

Fibonacci's sequence turns out to be just one special case of
sequences where you create new numbers by adding the previous two.

What if, instead of 1 1..., you start the sequence 1 3... ? And,
as with the Fibonacci sequence, you add the previous two terms
to get the next one. Starting with 1, 3 you get:

1 3 4 7 11 18 29 47....

This sequence is usually known as the Lucas sequence, after
Edouard Lucas, a Frenchman (so his name is pronounced Loo-
kah) who first explored it. It was Lucas who in 1878 actually
named the sequence that begins 1 1 2 3 5 8... the Fibonacci
sequence. So it took a mere 650 years for Fibonacci's name to
enter popular culture.

Put the Fibonacci sequence and the Lucas sequences
together, and the two seem to be completely unrelated:

FIBONACCI	1	1	2	3	5	8	13	21	34	55	89	144
LUCAS	1	3	4	7	11	18	29	47	76	123	199	322

Indeed, the two sequences are so different, mathematicians can prove that there is not a single member of the Fibonacci sequence that also appears in the Lucas sequence, apart from the numbers one and three.

And yet... Lucas discovered that the two series are intimately connected. Like Fibonacci numbers, Lucas numbers can be found in plants, for example certain types of cacti (though for reasons that aren't understood, Fibonacci numbers are much more common in plants than Lucas numbers).

In fact, the connections between Fibonacci numbers and Lucas numbers are much, much deeper than that. There are literally hundreds of different mathematical connections and formulae that have been discovered that link Fibonacci and Lucas numbers to each other. The simplest of them is perhaps the most astonishing of all.

If you divide each Lucas number by its corresponding Fibonacci number, something peculiar emerges:

LUCAS	1	3	4	7	11	18	29	47	76	123	199	322
FIB	1	1	2	3	5	8	13	21	34	55	89	144
LUCAS/FIB	1	3	2	2.333	2.2	2.25	2.231	2.238	2.235	2.236	2.236	2.236

As the numbers get bigger, their ratio seems to get stuck at a number that is roughly 2.236. Recognise it? We have met this number in the previous chapter. It's Fred, the square root of five, the length of the longest side of the triangle that replicates itself when you put five copies together. Fred has clearly got something to do with nature too.

There is, however, one final connection between Fibonacci and Lucas numbers that puts the others in the shade, and which has been responsible for catapulting these sequences to international fame. It was first discovered in the year 1609 by Johannes Kepler, the man usually remembered as the astronomer who discovered the true paths of planets around the sun.

The Golden Ratio

Kepler spotted that if you take any two consecutive terms in Fibonacci's sequence, their ratio is always roughly the same. As Kepler said: '...as 5 is to 8, so 8 to 13, practically, and as 8 is to 13 so is 13 to 21'. He noticed that as these numbers get bigger, the ratio settles on a number slightly larger than 1.6. For example $5/3 = 1.666$, $13/8 = 1.6125$ and so on.

He realised that if you chose a large enough number, this ratio becomes what we now know as 'The Golden Ratio', roughly 1.618 and symbolised by the Greek letter Phi. Now this was significant news, because at the time this ratio also went by the name of 'The Divine Proportion'.

The Golden Ratio was known to the Ancient Greeks, though they didn't attribute any divine properties to it. They merely knew that it was the ratio to be found in a particular rectangle that has a special property of creating miniature copies of itself if you chop a square off. Instructions on how to make a Golden Rectangle can be found on the page overleaf.

How to make a Golden Rectangle

Start by drawing a 2 x 2 square (two inches, two centimetres, it doesn't matter), extending the line that is the bottom of the square to the right. Then mark the centre of the base of the square with a blob:

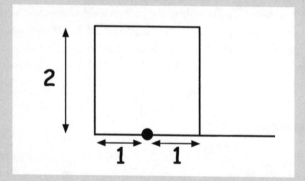

Put the point of a compass in the blob (E), and draw an arc of a circle that passes through the top right corner of the square (B).

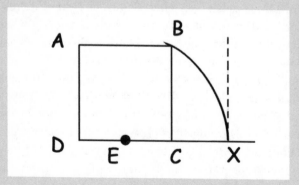

The point where the arc of the circle cuts the base line, X, marks the bottom right corner of the Golden Rectangle, ADXY.

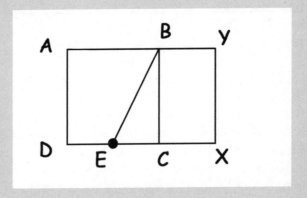

The ratio of the long side of the rectangle AY to the short side XY is Phi, 1.618... Incidentally, rectangle BCXY is also a Golden Rectangle, while triangle EBC is the Golden Triangle from Chapter 10. The length of EB is √5, or 'Fred'.

The Golden Ratio is not unique to Fibonacci, by the way. It appears in the Lucas sequence in just the same way. By the time you get to the tenth Lucas term, the ratio 123/76 is already almost spot on 1.618. Indeed, using Fibonacci's rabbit rules, you can start with any two numbers you like, call them Tom and Dick say, and remarkably after a few terms, Phi will always appear again.

For example, if Tom is 4 and Dick is 93, you get this:

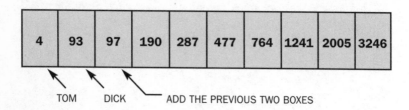

4	93	97	190	287	477	764	1241	2005	3246

TOM   DICK   ADD THE PREVIOUS TWO BOXES

By the time you get to the tenth box, the ratio of 3246/2005 is 1.619, already within 0.1 per cent of Phi.

It wasn't until the time of the Renaissance, when Ancient Greece became cool again, that folk like Leonardo da Vinci started to explore the Golden Ratio, and to find numbers and proportions in art and architecture that appeared to contain this very number. So when Kepler discovered the Golden Ratio in the Fibonacci sequence, its place in folklore was assured. Some even renamed it the Phi-bonacci sequence.

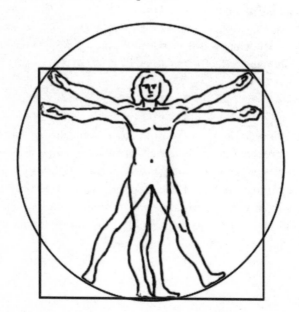

Debunking Da Vinci

At this point, familiar bells might start to ring for anyone who has read *The Da Vinci Code*. Early in the book, the hero character, Robert Langdon, is giving a lecture at Harvard, and begins to talk about Phi, the Golden Ratio. He says:

'Phi is generally considered the most beautiful number in the universe.'

Among other things, Langdon then goes on to claim that Phi is a 'fundamental building block in nature' and features in many of the most beautiful man-made structures too. His claims include the following list (repeated in other books). It may not surprise you, given the rather loose handling of 'facts' in *The Da Vinci Code*, that not all of them are actually true. Can you tell which of them are false?

1. If you divide the number of female bees by the number of male bees in any beehive in the world, the ratio is always Phi, 1.618.

2. In a nautilus sea shell, the ratio of each spiral's diameter to the next is Phi.

3. The human body is made up of building blocks whose proportional ratios always equal Phi.

4. Phi appears in numerous famous architectural structures, such as the pyramids of Egypt, the United Nations Building in New York and The Parthenon in Athens.

5. Compositions by Mozart, Beethoven, Schubert, Debussy and others feature Phi within their structure.

ANSWERS

1. False. The ratio is almost never even close to 1.618, not least because large numbers of male bees die off at certain times of year. Langdon has picked up on the Fibonacci bee pattern and completely misinterpreted it. Not very impressive for a Harvard professor.

2. False. The nautilus shape is a spiral that remains the same as you zoom out. The ratio between each spiral *could* be Phi, but in the real world it is usually about 1.3, a long way from Phi. It seems this myth was widely believed even by scientists until after *The Da Vinci Code* was published.

3. False for most people. Test it for yourself. Measure the following with a tape measure:

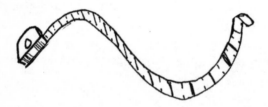

- The distance from your shoulder to your finger tip and your elbow to your finger tip;
- Your full height and the height of your navel above the floor;
- Your hip above the floor and your knee above the floor;
- The knuckle of any finger to the finger tip and the first joint to the finger tip.

Then work out the ratio of each of the pairs of measurements. In my case they are:

1.56 (shoulder/elbow)

1.70 (height/navel)
1.95 (hip/knee – though deciding where to measure the hip
was tricky)
1.85 (finger joints)

How about you? It's easy enough to find ratios around the
human body that are around one-and-a-half, but the claim that
they should all for some reason be Phi is completely spurious.

4. False. Just as with the human body, if you look hard enough at
the rectangles in a building, you will find some that are close to
being Golden Rectangles, but there are just as many that are close
to 1.5 or any other ratio you care to choose, too. Apparent connections between buildings and Phi are probably a coincidence, with
the exception of a couple of famous (and frankly rather ugly)
buildings that were deliberately designed using the ratio.

5. Might be true, but if so it's probably a fluke. Once again, if you
look hard enough in any pattern, you will find ratios around the
1.6 mark, but with the exception of Debussy (and some lesser
known composers) who deliberately built the ratio into their
music, once again it's a coincidence.

Dee dum dee dee dum

Which brings us back to Limericks, and their very catchy form.
Take a look at the Dees and Dums again, and count them. In the
first, second and third lines there are five Dees and three Dums.
In the third and fourth lines, it's three Dees and two Dums.

Dee Dum dee dee Dum dee dee Dum
Dee Dum dee dee Dum dee dee Dum
Dee Dum dee dee Dum
Dee Dum dee dee Dum
Dee Dum dee dee Dum dee dee Dum

The total number of Dees in the limerick is 21 and the total Dums is 13. Add up all the Dees and Dums and you get 34. *Every single number* we have counted is a Fibonacci number.

Unfortunately, while Fibonacci does work for the particular limerick at the start of the chapter, a look through any limerick collection might disappoint you. Very few limericks there fit the Fibonacci form, and to be honest, there are other patterns that work equally well.

Take for example this limerick, that goes back to my university days:

> *If you ask Mrs French for a yoghurt*
> *She'll be only too willing to flog it*
> *But at nineteen new pence*
> *It just doesn't make sense*
> *You can buy them much cheaper at Sainsb'ry's*

This time the Dums are Fibonacci numbers (3,3,2, 2 and 3 adding to 13), but the Dees are not – they are 7, 7, 4, 4 and 7, adding up to a total of 29.

So has this blown the whole theory away? Not quite. The Dees in the yoghurt limerick aren't Fibonacci numbers... but they are Lucas numbers. And the Lucas limerick number 29 divided by the Fibonacci number 13 is 2.231... remarkably close to Fred, the square root of five.

Fibonacci and the Golden Ratio are a part of mathematics where numbers meets numerology. But buried within the mysticism that this topic attracts, there is genuine mathematical beauty to be found in flowers, architecture, music, and perhaps even in a good limerick.

To infinity & beyond

You will need: some dominoes and a steady hand.

Buzz Lightyear's cry of 'To infinity and beyond' in the animated film *Toy Story* is one of the most enduring (and endearing) catchphrases in recent film history. Part of the charm of it is that it sums up Buzz's positive but rather dim way of thinking. Isn't the whole point about infinity that you can't go beyond it, however gung-ho your attitude is?

But perhaps Buzz Lightyear – or the scriptwriter behind him – is not as stupid as he sounds. The Infinity catchphrase might just have been a mathematician in the team having a private joke. Not only do mathematicians go to infinity all the time, they sometimes go beyond it too. And when you go to infinity, some strange things can happen.

Zeno's theory

Perhaps the first to grasp the strangeness of infinity was a man called Zeno, who wrote about a number of paradoxes some 2,500

years ago. The most famous is the story of the race between the fast-running Achilles and the plodding tortoise. He argued (with tongue slightly in cheek) that Achilles would never catch the tortoise. Suppose the tortoise has a 100 metre head start but Achilles runs ten times as fast.

By the time Achilles has run that 100 metres, the tortoise will have move forward by one metre. By the time Achilles runs the metre, the tortoise has progressed another centimetre. When Achilles has run that centimetre, the tortoise has nudged forward by a tenth of a millimetre. And this continues forever, each move forward by Achilles being matched by a smaller move forward by the tortoise. So Achilles can never overtake the tortoise.

Of course we all know that Achilles will overtake the tortoise, and he will do so at a point just over 101 metres from where he started, but Zeno's story is certainly enough to make you stop and think for a moment.

Zeno was also the first to ask what happens when you add one half to a quarter to an eighth, halving the number each time.

$\frac{1}{2} + \frac{1}{4} + \frac{1}{8} + \frac{1}{16}$...is a series that goes on forever – to infinity – yet it turns out that the answer is a finite whole number. The easiest way to demonstrate it is by drawing a picture. Starting with a whole square, we shade in a half, then a quarter, then an eighth, and so on, like this:

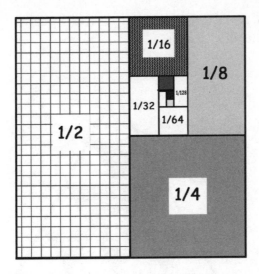

It is obvious now that the series will eventually fill the whole square, so the sum of the series is 1.

These sorts of infinite series crop up all over the place in mathematics, which makes them extremely important. Back in chapter 10, I posed the question: If you keep colouring in Sierpinski's gasket with black triangles, what proportion of the whole triangle ends up being coloured black?

The first black triangle takes out one quarter of the area...

Now each of the three white triangles has a quarter filled with a black triangle....

This leaves nine white triangles, each of which is one quarter filled with a black triangle. Keep filling the white triangles this way and you end up with a rather intimidating looking calculation: The total amount of black is $1/4 \times (1 + 3/4 + 3/4^2 + 3/4^3 + 3/4^4 \ldots)$

The series inside the brackets turns out to add to 4, and $1/4 \times 4 = 1$. In other words, if we keep on putting black triangles inside the white triangles, the whole triangle ends up as black. This may not seem that surprising, but there is a twist. Remember that the pattern of even numbers in Pascal's Triangle (page 140) is exactly the same as the pattern of black triangles in Sierpinski's triangle? What this means is, as you fill in more and more lines of Pascal, the proportion of even numbers heads towards 100 per cent, and the proportion of odd numbers heads to zero. This seems counter-intuitive – our old friend again! – since the edge of Pascal's Triangle is made up entirely of 1s, all of them odd numbers of course, so there are clearly an infinite number of odd numbers in Pascal's Triangle. The point is, there is an *infinitely bigger* number of even numbers.

Stacking books

So far when we've added up series of simple fractions they have ended up as a nice whole number. But that is not always the case. Take this series, for example, which is simply adding the fractions made from each of the counting numbers:

$$1 + \tfrac{1}{2} + \tfrac{1}{3} + \tfrac{1}{4} + \tfrac{1}{5} + \tfrac{1}{6} + \tfrac{1}{7} \ldots.$$

In this case as there is no end point, the sum keeps on growing. There is a nice way to prove it, too.

The first fraction is $\tfrac{1}{2}$

The next two are $\tfrac{1}{3} + \tfrac{1}{4}$.

$\tfrac{1}{4} + \tfrac{1}{4}$ is $\tfrac{1}{2}$, so since $\tfrac{1}{3}$ is bigger than $\tfrac{1}{4}$, $\tfrac{1}{3} + \tfrac{1}{4}$ must be bigger than $\tfrac{1}{2}$. (A quick check using decimals confirms it: 0.33333 + 0.25 = 0.583333, which is indeed bigger than 0.5.)

The next four fractions in the series are $\tfrac{1}{5} + \tfrac{1}{6} + \tfrac{1}{7} + \tfrac{1}{8}$. By the same reasoning, these four must add up to more than four eighths, and four eighths is $\tfrac{1}{2}$, so these four also add to more than $\tfrac{1}{2}$.

By grouping the fractions together, you can begin to see where this sum is heading:

$$1 + \tfrac{1}{2} + \tfrac{1}{3} + \tfrac{1}{4} + \tfrac{1}{5} + \tfrac{1}{6} + \tfrac{1}{7} + \tfrac{1}{8} + \tfrac{1}{9} + \tfrac{1}{10} + \tfrac{1}{11} + \tfrac{1}{12} + \tfrac{1}{13} + \tfrac{1}{14} + \tfrac{1}{15} + \tfrac{1}{16} \ldots.$$

This is $1 + \tfrac{1}{2}$ + more than $\tfrac{1}{2}$ + more than $\tfrac{1}{2}$ + more than $\tfrac{1}{2}$ ... forever, and if you keep adding halves forever you get to infinity – albeit half as quickly as if you keep adding ones together!

This rather abstract sum turns out to be connected to a very real problem sometimes encountered by librarians stacking books. At this point, you might like to try a little experiment, using some dominoes or other identical blocks such as hardback books.

Suppose you want to build a pile of dominoes and make it lean over. How far can you make your tower lean over before it begins to topple? What if you have 10 dominoes, or 100, or a million? Clearly there is some potential for the tower of dominoes to lean a bit without falling over. After all, the Tower of Pisa has been leaning over for centuries[6]. But, for example, is it possible for the top domino to stick out far enough that it is beyond the bottom domino?

One way to find out is to experiment, starting with the simplest case: two dominoes. A little trial and error – or a bit of engineering thinking – should convince you that it is possible for the top domino to protrude halfway over the bottom one without falling. The centre of gravity of the domino is in the middle of the block (indicated by a blob in the diagram), and as soon as that point moves beyond the supporting domino, the top domino will topple.

Half domino

---

[6] Actually the Leaning Tower is a little different, since its foundations are wedged into the ground, which helps to keep it upright.

What happens if you add a third domino? If it is placed on top of the other two and leans any further over than the half domino, the whole thing topples. However, it is possible to put it at the bottom of the stack, like this:

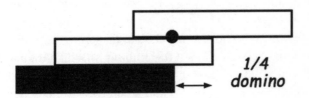

1/4 domino

The centre of gravity of the top two blocks is one quarter of a domino length away from the end, and as long as this centre of gravity does not extend beyond the bottom domino the pile will not topple.

As more dominos are added, the maths gets harder, but the idea is the same. To increase the amount of lean, keep adding dominos to the bottom, making sure that the centre of gravity of the top dominos does not extend beyond the end of the bottom domino. The pile with five dominos looks like this:

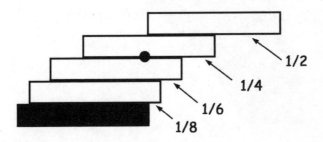

1/2
1/4
1/6
1/8

It turns out that the maximum lean of a pile of five dominoes is now $\frac{1}{2} + \frac{1}{4} + \frac{1}{6} + \frac{1}{8}$, which adds to very slightly more than one. In other words, the top domino is hanging beyond the edge of the bottom one – almost as if it is suspended in mid air. But if the top domino can stick out by more than one domino length after just five dominos, just how far is it possible to stick out with 10 or 20?

The clue is in a pattern of fractions that we've seen already. When four dominos are added the amount by which the top one can stick out is $\frac{1}{2} + \frac{1}{4} + \frac{1}{6} + \frac{1}{8}$. A fifth domino can stick out by a further $\frac{1}{10}$, a sixth by $\frac{1}{12}$ and so on. If this looks familiar, it's because it is exactly half of the series $1 + \frac{1}{2} + \frac{1}{3} + \frac{1}{4} + \frac{1}{5}$ ... And we already know that this series adds to infinity. Which means, so long as you add enough dominos, it is in theory possible to make the top domino protrude *by as many lengths as you wish*.

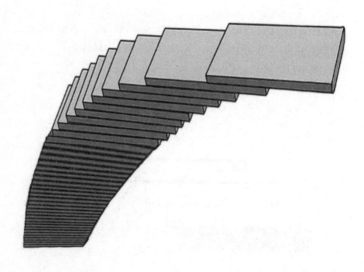

Be warned, however, the number of dominos that you need to make the tower extend outwards grows extremely fast. Only five dominos are needed for the top one to stick out by one domino length, but you need 32 dominos for it to be possible for the top one to protrude by two domino lengths, 228 for three domino lengths and nearly 2,000 for four. By the time you try to make the top domino stick out by 10 lengths, your pile of dominos will have extended beyond the moon, and the notion that your pile might be at risk of collapsing due to gravity becomes irrelevant – it will be floating!

Sums with debatable answers

We've seen that sometimes infinite sums add to nice whole numbers, and sometimes they add to infinity. Either way, it seems that at least they conclusively add up to something. But infinity is such a strange creature that it shouldn't surprise you that sometimes the answer to an infinite series isn't nearly so easy to nail down.

One of the weirdest results comes from looking at a very simple looking sum.

What is $1 - 1$? Zero of course

So what is $1 - 1 + 1 - 1 + 1$? Cancel out the positives and negatives and you should end with 1.

Those two were easy enough. But what happens if we go on forever? What is the answer to the sum that never stops, $1 - 1 + 1 - 1 + 1 - 1 + 1...$

Since each 1 is always followed by a –1, it seems reasonable to say that our infinite sum could just be written out in pairs, like this:

$$\boxed{1 - 1} + \boxed{1 - 1} + \boxed{1 - 1} + \boxed{1 - 1} + ...$$

And since this is just $0 + 0 + 0 + 0...$ the answer must be zero.

But there is no rule that says the brackets must go where we just put them. What if we do this instead:

1 - 1 + 1 - 1 + 1 - 1 + 1 - 1 + 1 ...

The sum has now become $1 + 0 + 0 + 0$... which is clearly going to end up as 1.

In other words, depending on how we look at it, the infinite sum $1 - 1 + 1 - 1$... **is either zero or 1**. But hang on, isn't maths supposed to be about getting to a single definite answer? Surely there can't be two answers? Let's try to resolve this little dispute.

Let's call the total sum that we are looking for *Bob*.

Bob = 1 - 1 + 1 - 1 + 1...

I'm now going to do a bit of manipulation by subtracting Bob from 1:

1 - Bob = 1 - (1 - 1 + 1 - 1 + 1 - ...)

The sum inside brackets is equal to Bob, and to demonstrate the weird thing about the Bob sum I'm going to need to remove those brackets. There is a rule for doing this when subtracting, which is to make all the positive numbers inside the brackets negative and the negative numbers positive. (You can either trust me on this, or demonstrate it to yourself using a simple sum, for example $10 - (2 + 3 - 1)$ which is $10 - 4$, which equals 6. Removing the brackets and reversing the signs of the numbers gives $10 - 2 - 3 + 1$, which is also 6.)

Removing the brackets from the Bob equation results in this:

1 - Bob = 1 - 1 + 1 - 1 + 1 - 1...

But now we are back where we started, because **1 - 1 + 1 - 1**... is Bob!

In other words this equation is saying that **1 – Bob = Bob**

What must Bob be, if 1 – Bob = Bob ? Bob can only be ½ – no other number works.

Pause for a moment and marvel at this. We've just done a little bit of algebra to demonstrate that the answer to the original sum 1 – 1 + 1 – 1 + … is neither 1 nor zero, but is instead ½! What a beautiful compromise, that answer is the average of the other two.

There are mathematicians who believe ½ is the best answer. Yet you might be feeling a little uneasy about this. There are no fractions involved in the sum, so why does one suddenly appear at the end? Welcome to another twist in the tale of infinity.

There is a similar problem known as Thomson's lamp. In this problem you start with a lamp that is switched on. After one minute you switch it off. After another 30 seconds you switch it on again, after 15 more seconds you switch it off, and you keep on switching it on and off again, halving the time interval between switches each time. The question is, after exactly two minutes, is the lamp on or off?

The maths behind Thomson's lamp seems similar to the 1 – 1 + 1 – 1 question, so perhaps this means that the lamp will not be ON or OFF but *half* on, glowing at half brightness. Are you convinced?

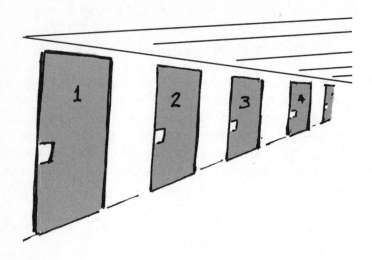

Beyond infinity

We started the chapter with Buzz Lightyear and his quest to go beyond infinity, and that is where we will end it.

One person who played with the idea of going beyond infinity was David Hilbert, a German mathematician with a very English sounding name. Hilbert imagined a hotel unlike any you will ever have come across. His hotel had an infinite number of rooms. Hilbert imagined what would happen if one day his hotel was completely full.

There was somebody in room 1, 2, 3, 4... up to any number you care to think of. At this point a customer knocked on the door, asking for a room. The manager thought for a second. 'Not a problem,' he said, 'I will find a room for you even though every room is full.' He sent a message to all of the guests, asking them each to move to a room whose number was exactly one higher than their current room. And this was possible, because in an infinite hotel, whatever numbered room you think of, there is always a number that is one higher.

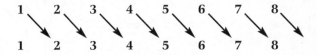

The result that every room was now filled, *except for Room 1*, and the new guest was checked into that room.

This is yet another demonstration that the maths of infinity is different from other maths. The normal symbol for infinity is:

The new guest at Hilbert's Hotel demonstrated that:

$$\infty = \infty + 1$$

The next day, something even more extraordinary happened. A coach filled with an *infinite number of passengers* turned up at the hotel. Remarkably, the manager did not bat an eyelid. 'I'll sort rooms out for you all,' he said. Again he sent a message to his guests (probably a bit disgruntled by now) asking each of them to move to a room whose number was double their current number. Every number has a double, so everyone was able to move.

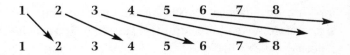

Notice how this time, all of the odd-numbered rooms became empty. Since there is an infinite number of odd numbers, this means that the manager was able to accommodate everybody on the coach.

In other words, Hilbert showed that.

$$\infty \: = \: \infty \times 2$$

This contradicts the arithmetic of normal numbers. And it sug-
gests that if you go beyond infinity, you still end up at infinity.

But that's not quite the case. Another German mathematician, Georg Cantor, was the first to establish that there are *different levels of infinity*. The infinities that Hilbert was playing with
in his hotel were infinities that you can 'count'. You can count
whole numbers, odd numbers, fractions (where the top and bottom are whole numbers) and plenty more. But there are bigger
infinities that you can't count. Lots of them. An infinite number.
For example – and you'll just have to trust me here – you can't
count 'irrational numbers', which are numbers that cannot be
represented as a fraction using whole numbers, like the square
root of 5 (Fred, whom we first encountered in Chapter 10) and
Phi (in Chapter 11).

For most of us, imagining infinity is hard enough. Imagining
infinities beyond infinities can be mind-blowing. More than one
mathematician has been driven to distraction in contemplating it.

It gives a hint of why maths is such a creative and imaginative subject. When people talk about drawing bar charts and
solving equations using a rote method and call it maths – well,
yes, that is a part of maths, and important too. But real maths,
the beautiful and exciting stuff, is about exploring, discovering,
making mistakes and occasionally encountering results that
are completely counter-intuitive. You don't find much of that in
your typical textbook.

I will leave the last word to David Hilbert. Hilbert once had a
student who was struggling with maths and had to leave the
course. Hilbert commented later:

*'He didn't have enough imagination to be a mathematician, but
now he's a poet and he's doing fine.'*